ATOMS, WHALES, AND RIVERS: GLOBAL ENVIRONMENTAL SECURITY AND INTERNATIONAL ORGANIZATION

ATOMS, WHALES, AND RIVERS: GLOBAL ENVIRONMENTAL SECURITY AND INTERNATIONAL ORGANIZATION

by
Peter J. Stoett
The University of British Columbia
Vancouver, Canada

NOVA SCIENCE PUBLISHERS, INC.

Art Director: Christopher Concannon
Graphics: Elenor Kallberg and Maria Ester Hawrys
Manuscript Coordinator: Roseann Pena
Book Production: Tammy Sauter, Benjamin Fung
and Michelle Lalo
Circulation: Irene Kwartiroff and Annette Hellinger

Library of Congress Cataloging–in–Publication Data
available upon request

ISBN 1-56072-235-5

6080 Jericho Turnpike, Suite 207
Commack, New York 11725
Tele. 516-499-3103 Fax 516-499-3146
E Mail Novasci1@aol.com

Printed in the United States of America

PREFACE

This book derives from what might be described as an intellectual intersection. An interest in politics collided with a deep concern for our planetary future, and the end result was a fascination with how conceptions of global security affect thinking about international organizations. I believe an effort is needed to demonstrate the potential utility of a concept of Global Environmental Security (GES) in this cross-disciplinary age of social science. For international relations scholars, the question is of direct relevance in the post-Cold War era as we struggle to redefine security in a manner more encompassing than traditional, military, and state-centric frameworks. For environmentalists, it is obvious that the lamentably predictable persistence of serious environmental problems worldwide warrants further research and theory.

This book is concerned primarily with the nexus between ecology, security, and international organization. The latter must be seen as a process generating institutions which are both promotional and constraining, and we must ask what effect the rise of environmental concerns has had on this activity. After examining the growing body of literature which questions traditional definitions of security, I arrive at a conceptual framework for GES. I then go on to examine in some detail three threats to GES which each capture different aspects of international environmental and political affairs: the spread of radioactivity and nuclear power, wildlife reduction in the commons, and regional resource conflicts. All these areas involve substantial interstate conflict and cooperation, the stuff from which international organization is made, and they have prompted the creation of regulatory regimes and intergovernmental institutions. The ones highlighted in this study include the International Atomic Energy Agency, the International Whaling Commission, and the Mekong Committee of Southeast Asia. This is not, however, a descriptive study of these bodies alone: we will go quite beyond them and into the realms of international political economy, environmentalism, and development. Using such varied case studies allows us to ask central questions about the contemporary international political system, such as regime leadership and the future of the nation-state in a GES-challenged world. I have learned much from the study of the politics surrounding atoms, whales, and rivers, and hope the reader will as well.

It should be noted that this book reflects research carried out in 1992-1993, and assumes no responsibility for events following that period. However, there is little substantial difference: nuclear power remains a contested international commodity, and fears of weapons proliferation have if anything increased; the plight of whales, and indeed all marine mammals, is still very uncertain and subject to complex political developments; and the Mekong continues to be threatened by the rapid economic growth in Southeast Asia and the plans of developers. More importantly, the broader question of what constitutes **global security** in this vulnerable age is still in need of serious discussion and input from environmentally-oriented analysts. GES is as important a topic as ever in the post-UNCED world, and we need to look critically as well at the important international institutions which will, in part, be responsible for its achievement.

I would like to thank the many individuals who have taken time to answer my questions related to this study (many are listed in the bibliography). I thank all the faculty members at Queen's University, Kingston, Ontario, Canada, where most of this book was prepared, most notably C. Pentland, D. Haglund, M. Hawes, and J. Lele. Special thanks go to R. Boardman for comments. Three colleagues, W. Cox, K. Ufori-Yeboah, and E. Laferrière, deserve special mention for long academic discussions. Any faults in this book are mine alone.

Writers need fairly extensive support networks as well, and I would like to acknowledge the professional and courteous aid of the secretarial staffs at both Queen's University and the University of British Columbia. As always, support from friends and family, especially my parents Fred and Irene, is appreciated. Finally, this book is dedicated to my partner-in-life, Betty McKee, and our daughter -- a relatively recent addition to the biosphere -- Alexandra Jane.

CONTENTS

CHAPTER FOUR

Atoms: Nuclear Trade, Weapons, and Discourse

CHAPTER FIVE

Whales: The Normative Transition of the International Whaling Commission

CHAPTER SIX

Rivers: Development, Geopolitics, and Environmental Alteration of the Mekong

LIST OF FIGURES

LIST OF ACRONYMS EMPLOYED IN THIS STUDY

ACUNS: Academic Council on the United Nations System.
ADB: Asian Development Bank.
AECL: Atomic Energy of Canada Ltd.
CITES: Convention on International Trade in Endangered Species of Wild Fauna and Flora.
EC: European Community.
EGAT: Electrical Generating Authority of Thailand.
FAO: Food and Agriculture Organization.
GES: Global Environmental Security.
IAEA: International Atomic Energy Agency.
IBRD: International Bank for Reconstruction and Development (World Bank).
ICCILMB: Interim Committee for Co-ordination of Investigations of the Lower Mekong Basin.
IDRC: Canadian International Development Research Centre.
INFCE: International Nuclear Fuel Cycle Evaluation.
IUCNNR: International Union for Conservation of Nature and Natural Resources.
IWC: International Whaling Commission.
MDC: Mekong Development Committee.
MDRN: Mekong Development Research Network.
NGO: Non-governmental organization.
SIPRI. Stockholm International Institute for Peach Research.
TERRA: Towards Ecological Recovery and Regional Alliances.
UNCED: United Nations Conference on Environment and Development.
UNEP: United Nations Environmental Programme.
UNESCAP: United Nations Economic and Social Commission for Asia and the Pacific.
UNESCO: United Nations Educational, Scientific and Cultural Organization.
UNFPA: United Nations Population Fund.
UNGA: United Nations General Assembly.
UNILC: United Nations International Law Commission.
UNTS: United Nations Treaty Series.
UNU: United Nations University.
USAID: United States Agency for International Development, Washington.
WCED: World Commission on Environment and Development.

CHAPTER ONE

INTRODUCTION: FORGING THE PROSCENIUM

"Actually no living organism exists on earth in a state of freedom. All organisms are connected indissolubly and uninterruptedly, first of all through nutrition and respiration, with the circumambient material and energetic medium. Outside they cannot exist in a natural condition."
V.I. Vernadsky, 1945[1]

"[P]ast expressions of neo-Malthusian ideas similar to the current limits-to-growth thesis have not led to the transcendence of narrow circumscribed loyalties ... national fears concerning over-population and insufficiency of raw materials have led to the most destructive and irrational of human impulses ... there is little evidence to suggest that mankind has advanced much beyond this level of jungle morality."
Robert Gilpin, 1981[2]

One of the 1993 Clinton appointees at the Department of Defense is officially titled the "Deputy Under Secretary for Environmental Security." The American National Intelligence Council, meanwhile, has been expanded to include an environmental spy-master. This immediately followed a year which saw the largest single international summit confer-

---

1 "The Biosphere and the Noosphere", *American Scientist*, 33 (January) 1945, 4. Vernadsky (1863-1945) was a Russian mineralogist who, amongst others, has been credited with developing the concept of the biosphere (discussed in Chapter Three of this study). To this Vernadsky added the idea of the noosphere, a psychological state of knowing and communicating, permeating the biosphere, an "anthropogenic era of geologic time." For discussion, see the introductory chapter in L. Caldwell's extensive text, *International Environmental Policy: Emergence and Dimensions*, Second Ed. (Durham, North Carolina: Duke University Press, 1990).

2 R. Gilpin, *War and Change in World Politics* (Cambridge: Cambridge University, 1981), 224.

ence ever, popularly dubbed The Earth Summit, in Rio de Janerio, Brazil. The United Nations Conference on Environment and Development (UNCED) involved, amongst other things, the adoption of the Rio Declaration and Agenda 21, which includes a commitment to create a Commission for Sustainable Development within the official UN structure.[3]

The ascendance of environmental problems, and by extension what some have termed the *ecological crisis*, in the arena of world politics has generated both optimistic and pessimistic responses. Environmental concerns such as acid rain, sea pollution, global warming, or ozone layer depletion offer physical reminders of the fact that the international system is a fundamentally interdependent one. Barring the invention and construction of rather large national biodomes, it is clear that the problems related to *global environmental security* (a term explained in greater detail in this study) cannot be solved on a unilateral basis.[4] As these problems are only increasing in scope and severity with the continued industrialization of human economic activity, the future promises more positive cognitive evolution at various levels of political decision-making, and it will become second-nature to turn to international diplomacy for assistance in implementing related policies aimed at reducing environmental risk. On another level of human interaction, special interest groups and\or social movements will continue to gain in importance and play a fundamental role in changing the more destructive intellectual perspectives that have taken us to this precarious point. This will occur in a smaller, but wiser world in which public policy and international organization will help solidify achieved balances between industry, environment, and population growth. The territorial division of the world becomes increasingly absurd; indeed it may one day be superseded by something else more conducive to the global management of resources and people.

---

[3] For an overview of the complex Preparatory process for UNCED, see S. Halpern, "The UNCED: Process and Documentation", *ACUNS Reports and Papers* (The Academic Council on the United Nations System), 1993; for post-Conference commentary see the special "Rio and After" issue of *International Journal*, XLVII:4 (Autumn, 1992). Within the following text, all excerpts from official UNCED documents are taken from a recent UN publication (also available on CD-ROM), *Earth Summit, Agenda 21: The UN Programme of Action From Rio*, Final Texts of Agreements Negotiated by Governments at the United Nations Conference on Environment and Development, 3-14 June 1992, Rio de Janeiro, Brazil (New York: UN, 1993).

[4] If the Biodome idea does not catch on, we might consider the proposal of Nikolai Agajanyan, a veteran of the former Soviet Union's space program. He recently told a conference for the International Congress of Physiological Sciences that people should seriously consider moving off Earth altogether: "mankind should live in space and the Earth should rest." *The Globe and Mail*, "Social Studies", August 4, 1992.

Needless to say, this is a rather optimistic outlook, and an unlikely one given the nature of the international political and economic system, both of which often seem founded on the principle of competition, not cooperation. One might, with Robert Gilpin, foresee the exacerbation of tensions as the end result of increasing environmental stress and scarcity. More people will compete for fewer resources, and the most aggressive unit of competition will remain the nation-state, always self-interested, always capable of resorting to warfare to pursue its ends. Rather than dilute the restrictive institution of national sovereignty, then, environmental problems will harden territorial lines, and this will be reflected in various painful episodes of the tragedy of the commons, in harsh immigration policies, and in the increased likelihood and occurrence of violent conflict. Not only will haves fight have-nots, but haves will fight haves and have-nots will fight have-nots. International anarchy and conflict will prevail, and in this social context the environment will become even less of a political priority than it already is. Decision-makers will as a result be preoccupied with a narrow vision of national security, one which can only increase insecurity in the long run, thus perpetuating the entire system.

As the opening quotations indicate, we live in a world where our interdependence is of the most fundamental nature, and yet where political divisions more often determine the fate of populations than their ecological solidarity. Contemporary humanity is located in a familiar, if always frightful location: at the crossroads. Familiar, because according to many understandings of history we are constantly making the future, if not in conditions of our own choosing; frightful, because the implications of forecasting and choosing routes have never been so ponderous. Different paths of action, taken in the past and present, will lead us towards either of the extreme scenarios envisioned above. Some of those paths, chosen by individuals and collectives, influenced by extant and evolving structures, will be found in the political realm of international organization. Here, thankfully, some trend toward the first scenario is visible. There is at present a plethora of environmental security measures instituted at the intergovernmental level, and though they hardly constitute a new world order, their breadth is impressive.[5]

[5] For two succinct (if already outdated) overviews, see P. Sands, "Enforcing Environmental Security", *Journal of International Affairs* 46 (Winter, 1993); and G. Plant, "Institutional and Legal Responses to Global Environmental Change", in I. Rowlands and M. Greene, eds., *Global Environmental Change and International Relations* (London: Macmillan and Millennium, 1992), 122-144. For official texts, see UNEP, *Selected Multilateral Treaties in the Field of the Environment*, I. Rummel-Balska and S. Osofo, eds., (Cambridge: Grotius, 1991).

This book, written largely during 1992-1993, will examine on a theoretical level the continuing evolution of the concept of security in world politics. The definition of security, as the literature review to follow will indicate, is slowly moving away from a fixation on military-related aspects of foreign policy. This does not necessarily mean the nation-state is disintegrating before our very eyes, but it does entail a broader perspective which may ultimately encourage a globalist ontology when it comes to viewing environmental problems and their political solutions. After a brief descriptive foray into the prevailing perspectives on world politics (pluralism, state-centricity, and globalism), the chapter will provide an examination of literature that challenges the traditional definition of security.

This will be followed by an attempt to explicate what, exactly, is meant by Global Environmental Security (GES), a concept becoming just popular enough to be in danger of also becoming meaningless. Environmental security refers essentially to the protection of habitat, which in today's global economy translates into an imperative need to protect the entire biosphere. Four essential complications will receive individual treatment: the idea of ecosystem health threats; political risk-oriented decision-making processes; the normative context, which suggests that GES is a fundamental human right; and the enticing as well as unavoidable questions of sovereignty, cooperation and conflict in the international system itself. What emerges is a unifying if complex concept, one which can be applied at a number of levels of analysis and which has both methodological and ethical implications. This will be followed by brief examples of threats to GES, including military conflict, transboundary pollution, resource scarcity, overpopulation, and climate change. Even these short ventures will demonstrate the impending complexity of the three case studies pursued in much more detail in this book.

Moving into an empirical mode, we will go on to look at three examples of how environmental threats have induced political decision-making in official international organizations and within the broader community[6] of actors involved in the issue-areas under investigation. The three issue-areas have been chosen because they represent a wide variety of problems and types of international organizations as well, it is freely admitted, as special interests of the author. The first is the question of atomic energy and the proliferation of nuclear weapons, which it is

[6] The term *community* is used generically here, and is hardly meant to connote a shared sense of perspective. In fact, environmental politics bring into focus sharp divisions in attitude and behavior.

argued should be seen as an environmental threat. This involves the trading regime in nuclear power, the non-proliferation regime, and the International Atomic Energy Agency (IAEA). The second is the loss of biodiversity, or rather the simplification of ecosystems due to species reduction, with a special look at the evolution of the whaling industry and the normative transition of the International Whaling Commission (IWC). The third is the environmental alteration created by major development projects and involves a regional international organization funded with multilateral development assistance, the Mekong Committee of Southeast Asia. We will deal, then, with a wide variety of physical entities, or morphic units: atoms, whales, rivers, and -- connecting them all -- people. But it should be obvious that we are in all cases dealing with the political behavior of the latter. Case studies include an historical depiction of the situation, an analysis of the current state of affairs, and limited projections concerning the future of the international organizations or regimes involved.

The spread of radioactivity is facilitated by both the horizontal and vertical proliferation of weapons and the export of nuclear power industry products. It is thus an issue encompassing both traditional military/security concerns, and the international political economy of trade in sensitive technology. The international organization which has evolved to deal with the threat posed by nuclear technology, in both its peaceful and military applications, is correspondingly complex, and as environmental and energy source concerns mount the question assumes greater urgency. Nuclear waste is not going away, and we are inundated with chilling news reports about weapons production–related atomic pollution in both the former superpowers. Due to the nation-state system itself, it is necessary to pursue as universal a regime as possible; indeed, the Non-Proliferation Treaty has the largest number of signatories in arms control history and, as we shall see, the IAEA is involved in promoting nuclear power in many diverse areas of the globe. For opponents of nuclear power, however, the latter aspect of the regime is more troubling than reassuring.

The politics of whaling, in contrast, have been much more limited; the threat of losing whale species has not quite compared to that of losing the human race through atomic warfare. Whaling involves a resource issue, if one can excuse the obvious and often criticized anthropocentrism required to label such majestic mammals "resources"; and it is also an issue of the global commons, since salt-water whales are migratory and in fact travel great distances on a yearly basis. Few issues have so sharply drawn lines between conservation and a certain industry as the whaling question. The number of nation-states taking an interest has been expanding as the latter issue has assumed greater importance with the serious depletion in whale populations and the rise of environmental consciousness, spurred by non-governmental actors, for which the whale has become a typical symbol. Due to the administrative structure of the IWC, any state can join the organization at any time, and it has been joined by a sufficient mass of non-whaling states to fundamentally change its normative bias. However, this has not translated into universal acceptance of the non-consumptive value of whales, as the continuing efforts to reorientate the Commission by resolute whaling nations such as Japan and Norway indicates. On the other hand, and especially since the recovery of limited stocks (such as the Minke), those nations quite correctly charge others with engaging in a form of what may be termed ethical or moral imperialism by insisting that whaling is in itself wrong.

Finally, the Mekong Committee[7] is by definition a regional organization, albeit one with a very wide mandate and international backing, and it is focused on a specific geographic area where interaction with the world economy has been rapidly increasing. The Committee's official members are the four riparian states of the Lower Mekong Basin (Laos, Thailand, Cambodia, and Vietnam); they may soon be joined by the upper Mekong states, China and Burma. However, the Committee is linked with the donor community by the development assistance it receives from Northern nations and the Asian Development Bank. The Committee functions in an exciting economic environment, arguably the fastest growing in the world economy today; and with it all the pain and dislocation brought on by rapid development. The Committee will constantly face related human rights questions and, increasingly, questions concerning the ecological sanity of some of the larger projects, such as dam construction, planned for the river. China's involvement adds a geopolitical twist. In short, the Mekong is an acid test of liberal institutional precepts; it may be a vehicle of cooperation in Southeast Asia, or it may prove -- as so many shared resources seem to -- to be the source of increased conflict. The effect of upriver alteration on downriver states, for example, is an obvious potential problem.

[7] The Committee has assumed a variety of names, including -- during Cambodia's temporary absence from 1975 to 1993, the Interim Committee for Co-ordination of Investigations of the Lower Mekong Basin. With the removal of obstacles to Cambodian participation, the Committee (which is probably, as we shall see, going to expand further) is no longer "Interim", and the Mekong Secretariat (the executive body in Bangkok, funded partly through ESCAP) simply refers to the Mekong Committee.

All three international organizations, and the broader regimes in which they function in an official capacity, have similarities. They are all related to resources, and to international resource sharing (be it technology, species, or geophysical entities). In other words, it has been impossible to deal with the relative issues without substantial international interaction, even in what have often been quite hostile political atmospheres. They were all formed in their present state shortly after the conclusion of that great transformative event, the Second World War. As such, the heavy American influence in their origins is hardly surprising, and seems to confirm assertions regarding the need for regime or even "hegemonic" leadership, at least at the formative stage. Each regime involves the participation, to a varying extent, of non-governmental organizations (NGOs), though success has been much more noticeable in the case of anti-whaling enthusiasts than the others. And they all, it may be argued, reflect the domination of a certain conception of human-nature ideology, a sense of engineered control over nature for human's benefit which is slowly evolving to reflect concerns with sustainability and, in some extreme cases, conservation.

There is more, from a critical perspective. All these regimes serve or at one point served the process of legitimizing potentially environmentally-destructive behavior. This is not a value judgement foisted upon those who worked diligently at these institutions, but is a statement reflecting this author's interpretation. The nuclear trading regime facilitates the spread of radioactivity and associated risks; the whaling commission once justified the slaughter, at clearly unsustainable yields, of Great Whales; and multilateral development programmes push for a path of industrialization that often leaves ecological chaos, and human rights abuses, in its wake. Regimes not only deter behavior, but promote it; the question that informs this book, then, is whether they are also susceptible to reform. This study will look at the many complex variables, principally in terms of events, ideas, and actors, that have come to influence these regimes as the environment becomes an issue of global concern.

Though they have much in common, the cases differ remarkably. It is clear that success, defined in terms of resultant reduction in threats to GES, comes easier in some areas than others. While all the international organizations listed above have been forced to adopt environmentalist concerns, this has had comparatively little impact in some. Only the IWC can be said to have evolved into an explicitly conservationist-oriented organization, and it is a strained one. Our ruminative conclusions will readdress the concept of security with reference to the discrepancies found throughout, and situate the findings in a broader theoretical discussion. We will also comment on the initial factors which comprised our framework for the study of GES, and engage in that most hazardous of intellectual occupations, modest predictions about the future of GES in general, and as it relates to international organization in particular. The pursuit of GES will in itself transform prevailing conceptions of security, but we should be modest in our expectations.

The tone of this study is one of very restrained optimism. So great are the problems associated with environmental degradation, so deep are the structures that perpetuate them, that a reflexive despair is common in the field today. Yet environmental concerns have generated much activity at both the diplomatic and, as importantly, transnational levels of interaction. This behavior helps forge an admittedly fragile but suggestive proscenium as we move into the next complicated century of human existence.[8]

[8] The proscenium originated in ancient Greek theatre; it was in essence a wall that formed a background for the actors.

CHAPTER TWO

THERE'S MUCH IN A WORD: SECURITY IN THE CONTEMPORARY WORLD SYSTEM[1]

"... a comprehensive approach to international and national security must transcend the traditional emphasis on military power and armed competition. The real sources of insecurity also encompass unsustainable development, and its effects can become intertwined with traditional forms of conflict in a manner that can extend and deepen the latter."
The Brundtland Report, 1987[2]

1. INTRODUCTION

The past two decades have seen a multitude of publications which challenge the traditional concept of national security. This literature often questions the logical primacy of the nation-state in modern security paradigms. If the world is as interdependent as analysts, politicians, and the managers of corporate globalization have been telling us, should we not focus conceptually on a level of security that is equally ambitious, and

[1] The original impetus for this literature review came from a paper written for Queen's *Centre for International Relations* in September of 1992 entitled "The Environmental Enlightenment: Security and Ecology." One version of this paper was subsequently presented to the Canadian Political Science Association, Ottawa, June 5, 1993; and another is published in *Coexistence: A Review of East-West and Development Issues* (Kluwer Academic, the Netherlands), 31:127-146, 1994. I am grateful to the Centre and its Director, David Haglund, for original funding.

[2] World Commission on Environment and Development, *Our Common Future* (Oxford: Oxford University Press, 1987), 290.

as encompassing? Of course, this is not an entirely novel question. From an ethical standpoint, many thinkers have been somewhat uncomfortable with the division of humanity the nation-state system represented, and with what some would claim were the resultant conflictual tendencies. On a policy level, there has been a sustained call for the establishment of *comprehensive*, as opposed to *collective*, security. This was articulated most famously by the 1982 Palme Commission, which stressed the importance of joint survival over what was at that time a fashionable, if terrifying, approach to the arms race, mutually assured destruction.[3]

Environmental problems have become fixtures in the arguments advanced for this attempt to broaden the definition of security not only in terms of the level of security that is attainable, but of what issue-areas, above and beyond relatively obvious military factors, should be considered integral to security. There is considerable controversy over this matter, since the concept of national security has a certain sacred ring to many military analysts, and a negative connotation to many environmentalists, or more broadly those identified with a peace research agenda. Some view the term as too specific to warrant expanded employment, others as too general to mean much subsequent to expansion.[4]

Indeed the word security, as Arnold Wolfers warned us many years ago, is at best ambiguous.[5] Security is the essential condition of human survival, and beyond that human prosperity. The definition of national security has begun to shift toward a more globalist perspective at least partly because of the intrinsic frustration, and empirical consequences, of the division of the world into nation-states. Once this has been established, it is argued, it becomes less novel, though certainly no

---

[3] Palme Commission on Disarmament and Security Issues, *A World at Peace: Common Security in the 21st Century* (Stockholm, 1989); see also *Concepts of Security: Report of the Secretary General* (New York: UN Department of Disarmament, 1986). For prescriptions for comprehensive security see R. Johansen, "Toward Post-Nuclear Global Security: An Overview", in Burns Weston, ed., *Alternative Security: Living Without Nuclear Deterrence* (Boulder: Westview, 1990), 220-260. Meanwhile, Norman Myers has published a book entitled *Ultimate Security: The Environmental Basis of Political Stability* (New York: Norton, 1993).

[4] For example, some participants at a recent Stanley Foundation conference "rejected the security concept for environmental issues ... *security* is an overused term in the Western political lexicon carrying a great deal of unwanted baggage." The Stanely Foundation, "Report of the 24th United Nations of the Next Decade Conference: Environmental Problems -- A Global Security Threat", June 18-23, 1989, Hamilton Parish, Bermuda, 15. For an excellent discussion see L. Brock, "Peace Through Parks: The Environment on the Peace Research Agenda", *Journal of Peace Research* 28:4 (1991), 407-423.

[5] "National Security as an Ambiguous Symbol", in A. Wolfers, *Discord and Collaboration: Essays in International Politics* (Baltimore: Johns Hopkins University, 1962).

less complicated, to think of security as a multifaceted concept, one which will never be viewed completely by anyone, but one of which all have some sight. Whether this will lead to the universal acceptance of, for example, environmental security as a foreign policy priority, is another question. As the case studies discussed in this book will indicate, this is more likely in some cases than others. But it is not without significance that there has been some perceptual alteration, predictably non-definitive, occurring in the security area. This chapter will examine this trend and offer a conceptual discussion of security in world politics today. But before looking at the security question in more detail, it will be helpful to examine an even broader conceptual landscape, that of world politics in general.

2. World Politics: Perspectives[6]

Though there are as many theories about international relations as there are theorists (and then some), it is beneficial to lay some groundwork in this early chapter if only to define terms which will be used throughout this study. With this modest task in mind we may proceed to describe three prevailing perspectives of world politics. Aware of the stern and sage warning from Albert Hirschman,[7] we will simplify them according to their main tenets and assumptions, and describe them in terms of a theoretical trichotomy: pluralism, or liberal institutionalism; state-centricity; and two strands of what has been popularly termed globalism. We should, however, add the caveat which usually proceeds such gross simplifications of what is actually a rich and complex theoretical terrain. We do not wish to reduce the study of world politics to three competing perspectives, but rather use them to indicate rough demarcations in published thinking about the subject. Such demarcations are permeable, fluid, and subject to dialectical alterations. Indeed, the intro-

[6] Many of the ideas presented here were first treated in a paper delivered by the author to the *Fifth Conference on the United Nations and Peace,* University of Victoria, B.C., February 1992, entitled "Competing Perspectives of World Politics: Environmental Crises and Global Security."

[7] Hirschman has referred to the "search for paradigms as a hindrance to understanding"; "a disease at least as prevalent and debilitating ... as the spread of mindless number work in the social sciences." *A Bias For Hope: Essays on Development and Latin America* (New Haven: Yale University Press, 1971), 342-360.

duction of ecological questions to the field, which has been slowly taking place over the last two decades, illustrates this conceptual dynamism.

I. PLURALISM

There is little doubt that, overall, what we here term the pluralist model of world society - an anarchical system in which cooperation is firmly institutionalized - dominates current Western thinking in global environmental politics, both at the academic and policy levels. Such thought, in fact, has a surprisingly "long pedigree, going back to at least utilitarianism, the technocratic politics of Saint-Simon, the 19th-century Manchester School, and Fabian socialism."[8] Other terms associated with its contemporary variations include Grotian, liberal institutionalism, and interdependency (some would include "neorealism" as well.) A faith in the essentially compatible nature of humanity permeates such theory, which is most often rooted in liberal assumptions about trade, state/society distinctions, and human ingenuity and rationality ("enlightened self-interest").

At an extreme edge, pluralism is most complete in scenarios of world government. In this context the world could truly consist of a democratic pluralist society, complete with some sort of enforcement mechanism for commonly accepted rules. Outside of this futuristic realm, thinking about such a grand design has been much more restrained. Integration theorists have written of the gradual establishment of supranational governments, be they along federal or, less bonding, confederal lines. *Functionalism,* a "normative political theory from a left leaning liberal" in the person of David Mitrany, emerged as a branch of such thinking following World War Two.[9] Functionalists envisioned such integration as a process arising out of technical cooperation amongst nation-states, and the behavioralist-era neofunctionalists stressed the role of

---

[8] C. Pentland, "Integration, Interdependence, and Institutions: Approaches to International Order", in D. Haglund and M. Hawes, ed., *World Politics: Power, Interdependence, and Dependence* (Toronto: Harcourt Brace Jovanovich, 1990), 173-196, 174.

[9] D. Long, "From World Peace to International Organization: The Development of Functionalism as a Theory of International Relations", paper presented to the Canadian Political Science Association Annual Meeting, Carleton University, Ottawa, 6-8 June 1993. Mitrany's classic text is *A Working Peace System: An Argument for the Functional Development of International Organisation* (London: RIIA, 1943). Mitrany would be critical of political integration on a regional level, preferring the eventual establishment of a post nation-state world.

mutual self interest in the construction of institutions whose success would "spillover" into other areas of interaction.[10] In the development of enlarged political community, then, form should follow function. The contemporary utility of the nation-state is challenged by this thinking, which oscillates in terms of relevance and fashionability, usually according to the political state of the European Community (now known as the European Union).

The ecological crisis[11] does create ample room for functionalist thinking. This may be manifested at the institution-building level: the granting of authority to supranational entities in which scientific consensus determines policy. This would include for example hopeful calls for making the United Nations Environmental Programme (UNEP) a much stronger institution in the hope that it could break down political boundaries.[12] Over thirty nations signed the non-binding *Hague Declaration* in March of 1990, which calls for the establishment of a new world authority to prevent further global warming and advocates the use of the International Court of Justice to arbitrate disputes.[13] However, the Hague Declaration remains little more than the result of a foreign policy excursion, superseded by the less demanding UNCED agenda (significantly, the United States and what was then the Soviet Union did not even attend

---

[10] Most famously, see E. Haas, *Beyond the Nation-State* (Stanford: Stanford University Press, 1964); and also A. Groom and P. Taylor, eds., *Functionalism: Theory and Practice in International Relations* (London: University of London, 1975).

[11] This is a generic term, used to denote the multiple environmental problems faced by contemporary earth as a whole. It has become common to indicate their intertwined nature by using the singular term *crisis*. Protests that this implies a bias toward globalist or pluralist perspectives are perceptive. Ernst Haas, discussing "eco-reformers", makes mention of a similar concept, that of *the problematique*: "the problem of all the problems, not merely the sum of the problems of pollution, war, famine, alienation, resource depletion, urban crowding, and exploitation of the Third World by the First. It is a systemic construct that assumes causal connections among these problems, connections that amplify the disturbance in the metasystem." "Words Can Hurt You: or, Who Said What to Whom About Regimes", in S. Krasner, ed., *International Regimes* (Ithaca: Cornell University, 1983), 23-59, 39. Haas makes reference to Hasan Ozbekhan, "The Predicament of Mankind", in C. W. Churchman and R.O. Mason, eds., *World Modelling: A Dialogue*, vol. 2 (New York: American Elsevier, 1976), 16.

[12] I have made similar appeals in the past, from what can only be described as a rather less complicated perspective than that which permeates the present study. See "Of Opportunity and Necessity: The United Nations Environmental Programme and the Functionalist Approach", *International Insights* (Dalhousie University, Halifax), 6:1 (Spring 1990), 47-54. We might note that Chapter 38:22 of Agenda 21 (UNCED) calls for the UNEP to strengthen its "catalytic role in stimulating and promoting environmental activities and considerations throughout the UN system." This is a more realistic prescription.

[13] See UN Document A/44/340-E/1989/120, annex.

the Hague Ministerial Conference). Meanwhile the evidence in terms of a neofunctionalist trend in the continuing evolution of the European Community, often considered the most fertile proving ground for what is admittedly a rather Eurocentric theoretical proposition, seems rather bleak.[14]

What remains resonant from the functionalist literature is the notion that the modern state is not equipped to deal with the daily needs of contemporary citizens. In a certain sense, there is little doubt that a growing number of states, due to ecological and other problems, are totally inadequate in terms of governance. Lynton Caldwell writes of the potential spread of what he terms *socio-ecological insolvency*, wherein "a state has exhausted its material means of self-support and no longer provides to its people the elementary services of government."[15] A stark contemporary example of such a state was war-torn and soil-eroded Somalia in the early 1990s; long-term prospects include most states, especially those with unbearable population pressures (Bangladesh, China). And functionalism, stressing technocratic social democracy, foresaw the formation of groups of international scientists acting in concert to influence policy, commonly referred to as "epistemic communities."[16] However neither of these factors points to an evolutionary end of the sovereign state system, or to the conflicts between states.

Less ambitious pluralism attempts to come to terms with the inconvenient fact that the battered but staunch nation-state just won't go away. Ideas about pluralism in world politics are currently manifested in academic inquiry by what has been popularly labelled *regime analysis*, stemming from mainstream academia's mid-1970s preoccupation with the concept of interdependence in world politics. In order to cope with this interdependence -- defined primarily in economic terms, though this is slowly changing to reflect ecological matters -- nations form regimes, defined succinctly enough by Stephen Krasner as sets of principles, norms, rules, and decision-making procedures around which actors' expectations converge. Regimes are often regarded as "intervening variables": they do not change the fundamental structures of political power, but they may

---

[14] See in particular M. Huelshoff and T. Pfeiffer, "Environmental Policy in the EC: Neo-functionalist Sovereignty Transfer or Neo-realist Gate-keeping?" *International Journal* XLVII:1 (1992), 136-158.

[15] L. Caldwell, *International Environmental Policy: Emergence and Dimensions*, Second Edition (Durham: Duke University Press, 1990), 328.

[16] I am grateful to David Long for emphasizing this point. On epistemic communities, the latest fashion in regime analysis, see P. Haas, "Introduction: Epistemic Communities and International Policy Coordination", *International Organization* 46:1 (1992), 1-35.

influence the ultimate outcome of behavior emanating from the international system.[17] Of course, one may be more or less enthusiastic about just how "intervening" these variables are. This doesn't always seem to matter: in its current usage, the term regime has acquired fantastic flexibility. A loose definition of what, exactly, constitutes a regime -- and a tight definition is unnecessarily constraining -- leads to the conclusion that most areas of international collaboration do so, and whether or not some hegemonic leader provides the "public good" of leadership. What were once functionalist projects, for example, have become regimes.[18]

Yet the appeal of regime analysis at this stage in the development of global environmental politics is self-evident. Though it may be considered a shallow, even cosmetic perspective by those who prefer grand theory, one of the strengths of such an approach is that it facilitates prescriptive analysis (in other words, it calls for the establishment of more, or better, regimes). Some scepticism is warranted, however. Regimes are in effect a one-worded and relatively fresh way of saying international organization. Within regimes one will find actual organizations set up by governments and non-governmental actors, which if nothing else have a separate identity from the individual actors which have established them. In the more complex sense, there are social constructs which influence these spaces of interaction, the normative context of behavior. Identification of the latter can only be an imprecise enterprise, perhaps largely determined by the intellectual perspective of the observer. Oran Young's differentiation between institutions ("social practices consisting of easily recognised roles coupled with clusters of rules or conventions governing relations among the occupants of those roles") and organisations (material, extant entities which possess legal sovereignty and physical artifacts, such as office buildings) is helpful.[19] But the difficulty of creating a universally applicable taxonomy of regimes should be self-evident.

---

[17] For standard texts see S. Krasner, "Structural Causes and Regime Consequences: Regimes as Intervening Variables", in S. Krasner, ed., *International Regimes* (Ithaca, NY: Cornell University Press, 1983), 1-22; R. Keohane, *After Hegemony: Cooperation and Discord in the World Political Economy* (Princeton, NJ: Princeton University Press, 1984); M. Zacher, "Toward a Theory of International Regimes", *Journal of International Affairs* 44:1 (1990), 139-158; O. Young, "The Politics of International Regime Formation: Managing Natural Resources and the Environment", *International Organization* 43:3 (1989), 349-375.

[18] For example, see P. Sands, "EC Environmental Law: The Evolution of a Regional Regime of International Environmental Protection", *The Yale Law Journal* 100:8 (1991), 2511-2523.

[19] O. Young, *International Cooperation: Building Regimes for Natural Resources and the Environment* (Ithaca: Cornell University Press, 1989), 32.

Liberal institutionalists can point with some satisfaction to a growing network of regimes devoted to environmental problems at the intergovernmental level. This includes the formal expression of agreements, conventions, and declarations, and the creation of new organizations. Just as importantly, as this study will show, the participants in extant formal arrangements and international organizations are adopting environmental concerns, and newfound articulations of them, as policy priorities. One example is the renegotiated Antarctic Treaty signed in 1991, which prohibits mining on Antarctica for the next fifty years. There are many others, including the various issues explored in this book (nuclear power, wildlife conservation, development projects). Already it is obvious that "UN policymaking in ecosystem issue areas [has] focused primarily on the creation of international regimes [which] frequently required new institutions for implementation and monitoring."[20] Hillary French found in 1992 "more than 170 environmental treaties concerning subjects of shared concern: acid rain, ocean pollution, endangered species protection, hazardous wastes export, and the preservation of Antarctica, as well as ozone depletion."[21] Jim MacNeil also acknowledges that "extensive institutional structures" exist, but reminds us they lack "political strength, cohesion, depth, and financial capacity."[22] The broader question, much more difficult to answer yet equally important, is whether all of this institutional activity is changing the normative context in which it operates. It is at this point that the state-centric and globalist perspectives offer competing critical insights, but a few extra words may be said first.

Though they permeate the literature, it is quite desirable to put some of the more optimistic liberal assumptions aside.[23] Regimes, if we

---

[20] D. Puchala, "The UN and Ecosystem Issues: Institutionalizing the Global Interest", in L. Finkelstein, ed., *Politics in the United Nations System* (Durham: Duke University Press, 1988), 214-242, 218.

[21] H. French, "After the Earth Summit: The Future of Environmental Governance", *World Watch Paper* 107 (March 1992), 6.

[22] J. MacNeil, et. al., *Beyond Interdependence: The Meshing of the World's Economy and the Earth's Ecology* (New York: Oxford University, 1991), 120.

[23] James Keeley, calling for more sophisticated regime analysis, writes that "[o]ne fundamental problem of regime theory ... is that it has been placed in a continuum between a simplistic realism and an apologetic and hopeful liberalism. In such a contest, over the long term one would expect the liberals to be 'mugged by reality'". "The Latest Wave: A Critical Review of Regime Literature" in Haglund and Hawes, *op. cit.*, 553-569. A similar call for advanced work in this area, which stresses the need to link domestic societies with regimes, is an article by S. Haggard and B. Simmons, "Theories of International Regimes", *International Organization* 41(3) 1987, 491-517. This is an avenue of great potential for students of environmental policy.

do view them as causal agents in the international system, cause problems as well as provide possible avenues toward solutions. One need not be blind to this less optimistic perspective on regimes, one with some grounding in so-called postmodernist literature. While regimes articulate agendas for action, they restrict them as well, by defining what is acceptable discourse. The chapters that follow will demonstrate this is the case; although, since no regime is an island, there are always extra exogenous actors and structural variables that challenge this ideational hegemony. And, as our studies will also make clear, regimes are principal variables in the process of legitimization. This may be negative (delegitimization), involving redefining a certain type of behavior as illegitimate and proscribing it. These have been specifically referred to as *global prohibition regimes*: they are guided by norms which "strictly circumscribe the conditions under which states can participate in and authorize these activities and proscribe all involvement by nonstate actors."[24] At the same time, regimes have a corresponding positive function, to legitimize behavior which is taking place. This could include, for example, behavior that is arguably hazardous to environmental health, such as the spread of nuclear power, or the unsustainable killing of wildlife. The tendency to equate regime formation with a progressive evolution in world affairs overlooks the dual nature and, at a policy level, the possible manipulation of such institutions.

As importantly, the role non-governmental actors play in international organization is essential. Surely one of the greatest challenges made to conventional international political theory by the ecological crisis stems from the necessary infusion of non-governmental participants in any evaluation of the situation, including industrial, societal, and even transnational actors. As Simon Dalby writes, "focusing on states as the [sole] providers of security limits the possibilities of active intervention by NGOs (nongovernmental organizations), social movements, and individuals in the process of international politics."[25]

---

[24] E. Nadelmann, "Global Prohibition Regimes: the Evolution of Norms in International Society", *International Organization*, 44:4 (1990), 481-526, 481.

[25] "Security, Modernity, Ecology: The Dilemmas of Post-Cold War Security Discourse", *Alternatives*, 17 (1992), 95-134, 102.

II. STATE-CENTRICITY

What is probably the oldest line of thought about international relations has been based on the assumption of state-centricity; it has been called "realism", also known as statist or mercantilist and, even, a Machiavellian perspective.[26] Realists are fond of reminding us that their particular outlook, which centers around a dichotomy between the concepts of anarchy and order, dates back at least to Thucydides' analysis of the Peloponnesian War (430 B.C). The basic unit of analysis remains the state (be it a city-state or a nation-state) or the system of states; and the basic problem of international relations remains war. States, due to the inevitable "security dilemma" created by individual attempts to enhance security, are preoccupied with consequently increased national security (defined especially in military terms) considerations. Thus state-centricity is assumed here to adhere to an intellectual military essentialism in questions related to security.

As Robert Gilpin writes, "realism holds that through calculations of power and national interest statesmen can create order out of anarchy and thereby [at best] moderate the inevitable conflicts of autonomous, self-centered, and competitive states."[27] That these conflicts are regarded as inevitable reflects a rather pessimistic, "Hobbesian" view of human nature based on the imperative of aggressive power-maximization; foreign policy is thus broadly defined by Hans Morgenthau as a "continuing effort to maintain and to increase the power of one's own nation and to keep in check or reduce the power of other nations."[28] The collective experience of impending doom -- what many pluralists and globalists see as a possible cataclysm toward a better world -- is thus viewed with frank scepticism. States, though capable of co-operation, will jealously guard their legal sovereignty, and regimes are little more than epiphenomena.[29]

In terms of the traditional analysis of national security as a foreign policy problem, it is quite understandable that state-centricity has held a

---

[26] The use of the word "realism" to describe a perspective on world politics could be one of the greatest semantic coups in academic history. All those rejecting realism as a paradigm could be lumped into the category of "idealists." The connotations are obvious, and misleading. Such a simplistic split is unfair to both realism and non-realist theory.

[27] *War and Change in World Politics* (New York: Cambridge University Press, 1981), 226.

[28] *Politics Among Nations: The Struggle for Power and Peace* (New York: Knopf, 1948), 223.

[29] For realist critiques of regime analysis, see especially S. Strange, "*Cave! hic dragones:* A Critique of Regime Analysis", in S. Krasner, ed., op. cit., 337-354; and J. Grieco, "Anarchy and the Limits of Cooperation: A Realist Critique of the Newest Liberal Institutionalism", *International Organization* 42 (Summer 1988), 485-507.

firm ontological grip on prevalent perspectives. This is especially so since the Cold War dominated so much scholarly and policy oriented activity in the most academically prolific nations -- the United States and Western Europe. This reflected an ethnocentric preoccupation with a bipolar structure of world politics. Yet a world free of the shackles of superpower rivalry, as balance of power theorists and daily newspapers persistently remind us, is by no means an inherently stable one. Academically, the Cold War acted largely as a perceptual fog, which (with well-funded consistency) clouded over many of the problems we now face. In particular, security was defined in such a parochial manner that new conceptions are just now making their theoretical debut in mainstream literature.[30] Defining security within what is cursorily known as the "realist paradigm" exaggerated the amount of order the superpowers could impose on this world of complex political chaos. Indeed, one can even discern a certain creeping nostalgia for the good old bad days of the limited certainty provided by the Cold War.

As the impermeable state becomes a more difficult concept to consider plausibly, variants on realism have evolved. *Neorealism* accepts many of the claims of liberal institutionalists; *structural* realism continues to stress basic power relations and position. This conceptual jostling, beyond its often coerced relevance to graduate students of the discipline, serves to prevent the removal of the state from the center of analysis (pluralism, especially regime analysis, may be similarly characterized). Biospheric problems present an immediate challenge to state centricity, defying the boundaries upon which it is based. However, and though it is certainly time to move beyond state-centricity in environmental analyses, it is another matter to suggest that governments will move visibly beyond national interest as a canonized guiding principle, or dismiss their own sovereignty as inconsequential. The weakness of constraints on anarchy represented by international law remains evident in any issue-area, especially one as contemporary as environmental security. It is not only premature, it would be delinquent, to disregard the contributions of realism at this stage; this is especially so given the fact that governments around the globe have moved quickly to capture the dominant discourse on envi-

---

[30] For an absorbing and recent Foucaldian analysis of the American construction of the concept of security, see D. Campbell's *Writing Security: United States Foreign Policy and the Politics of Identity* (Minneapolis: University of Minnesota Press, 1992). And for a critical discussion on "the extent to which conventional accounts of security depend on certain assumptions embodied in the principle of state sovereignty" see R. Walker, "Security, Sovereignty, and the Challenge of World Politics", *Alternatives* 15 (1990), 3-27.

ronmental futures. And geopolitical literature linking natural resources and national strategy has a promising future.[31]

There is plenty of room for state-centric power analysis within the scope of global environmental security, but power has taken on a new form. India, China and Brazil all present a threat to future generations if they follow the development paths of the west, and as such the populous southern states wield considerable leverage. This North-South dimension will dominate related diplomacy in the future. This was evident in the negotiations following the Montreal protocol, which established a complicated ozone-protection regime contingent upon the inclusion of a technology-transfer clause. The very threat to biospheric integrity posed by the future use of chlorofluorocarbons by heavily populated southern states lent the south what may be epiphenomenal, but by no means ephemeral, bargaining power.[32] Yet the ecological crisis does confront the logic of the model of a zero-sum world of states (where one states' gain is necessarily another's loss) that previously dominated the discipline of international relations. Further, the North-South divide is assuming greater importance than the overall division of the world into territorial states.[33] In fact, any case study of global environmental security, including those which compose this study, invariably deals to some degree with the North-South dynamic. It has become as contextual as the anarchical system itself, although there are problems with taking the North-South divide as primary -- which transnational class analysis reveals.

And finally there is the normative, or rather ethical, element of state-centricity that many environmentalists and others find disturbing. Though it is unfair to assumedly link the pessimistic Neo-Malthusian Garrett Hardin's "lifeboat ethics" with realism (or, perhaps, vice-versa), one cannot help but see at least a superficial linkage.[34]

---

[31] See R. Lipshutz, *When Nations Clash: Raw Materials, Ideology, and Foreign Policy* (New York: Ballinger, 1989); I. Lesser, *Resources and Strategy* (London: Macmillan, 1989); D. Haglund, "The New Geopolitics of Minerals: An Inquiry into the Changing International Significance of Strategic Minerals", in same, ed., *The New Geopolitics of Strategic Minerals: Canada and International Resource Trade* (Vancouver: University of British Columbia, 1989).

[32] See R. Benedick, *Ozone Diplomacy: New Directions in Safeguarding the Planet* (Cambridge: Harvard University Press, 1991), 76-116.

[33] K. Ramakrishna, "North-South Issues, the Common Heritage of Mankind and Global Environmental Change", in I. Rowlands and M. Greene, ed., *Global Environmental Change and International Relations* (London: Macmillan and *Millennium*, 1992), 145-168; L. Lunde, "North/South and Global Warming -- Conflict or Cooperation?", *Bulletin of Peace Proposals* 22:2 (1991), 199-210.

[34] See Hardin's 1968 article, "The Tragedy of the Commons", *Science* 162: 1243-1248; and "Living on a Lifeboat" in the October 1974 issue of *BioScience* (24:10, 561-568), in which he

III. GLOBALISM, STRUCTURAL AND NORMATIVE

It is for the sake of procrustean taxonomy that we label quite divergent theoretical paths "globalist", but there are two essential orientations here, both of which hold that the only proper level of political analysis is global. Neither, it should be stressed at this point, refers to the process of globalization as envisioned by the neoliberal economic agenda. One, which is largely found in the literature on international political economy and has a long intellectual history which includes marxist studies of imperialism and dependency, may be termed *radical structuralism*. Its contemporary proponents argue that the capitalist world system is based upon a core-periphery relationship between North and South; the state system is superimposed upon this material base. Immanuel Wallerstein is the principal modern architect of this approach, which stresses the dialectical development of a hegemonic world economy and class analysis across nation-states, but there is a wealth of literature with this orientation.[35] The central objective is, in words of Herbert Addo, "to understand the phenomena of exploitation, so that the forces and structures which give rise to and sustain it can be influenced in directions that will make them ineffective."[36] This is a tall order. We might take Sarah Brown's definition, that the analysis of international relations is the study of "the identification and explanation of social stratification and of inequality as structured at the level of global relations."[37] This is equally ambitious, introducing issues such as gender and race, and is as charged with a fundamental concern for social justice in a transnational context. However, such an agenda leads to complications as well.

---

writes: "For the metaphor of a spaceship to be correct the aggregate of people on board would have to be under unitary sovereign control." For an eloquent response to Hardin's line of reasoning, see M. Soroos, "The Commons and Lifeboat as Guides for International Ecological Policy", *International Studies Quarterly*, 21:4 (1977), 647-674.

[35] It should be noted that there is much variation within this field. See, for example, I. Wallerstein, *The Capitalist World Economy* (London: Cambridge University Press, 1979), and *The Politics of the World Economy* (Cambridge: Cambridge University Press, 1984); C. Chase-Dunn and R. Rubinson, "Toward a Structural Perspective on the World System", *Politics and Society* 7:4 (1977), 453-76; A. Brewer, *Marxist Thought on Imperialism: A Critical Survey* (Boston: Routledge and Kegan Paul,1980); R. Cox, *Production, Power, and World Order* (New York: Columbia University Press, 1987).

[36] H. Addo, *Transforming the World Economy? Nine Critical Essays on the New International Order* (Toronto: Hodder and Stoughton, 1984), 246.

[37] S. Brown, "Feminism, International Theory, and the International Relations of Gender Inequality", *Millennium* 17 (1988), 461.

If one is convinced that capitalism is the ultimate cause of environmental destruction, incremental reform means little. For example, Samir Amin, a leading student of capitalist imperialism, believes that the currently vogue Green movement, assumed by some to have global transformational potential, appears on the political scene much like "other forms of religious fundamentalism," not as a movement offering real "solutions to the crises, but rather as symptoms of it."[38] The real crisis results from the internal contradictions of capitalism on a world scale. The theoretical proposition may be sound; one need not look far to see the ravages of capitalist accumulation in the world, and much ecological destruction is the direct result of the separation of peripheral people from their land, and the drive for profits emanating from the core. But the prescription for world socialism remains as distant a dream as ever. This does not deny the potential for global transformation in such a direction.

The second approach with global aspirations, one with many overlappings with pluralism, we may term *normative globalism*. There is, among analysts of various political stripes, a belief that human society is becoming closer as the world shrinks, and that we are on our way to some type of global ethical framework regarding questions of social justice, peace, and environment, perhaps even a "planetary consciousness".[39] This is a crude representation of some very sophisticated and diverse work, but there are certain similarities. Again, there is a rejection of the state-centric paradigm, to which the pluralists described above are critically assumed to belong. Accent is placed on the role of non-governmental actors, in particular mass, post-modern movements, including feminism and post-materialism, as agents of social change (whereas radical structuralists will generally look only to classes for this impetus). As Robert Walker writes, "large-scale transformation [and ecological change is certainly an example] always brings forth new political actors and new forms of political action. Social movements ... have always arisen to express opposition to dominant power structures as well as to respond to structural dislocation."[40] The answer to the problems with which international relations theory struggles is some sort of unprescribed world

[38] S. Amin, "A Propos the 'Green' Movements", trans. by P. Burgess and C. Blower, in H. Addo, et. al., *Development as Social Transformation: Reflections on the Global Problematique* (London: Hodder and Stoughton, with the United Nation University, 1983), 270-281; 280.

[39] The term has been employed by Richard Falk. See the "Introduction", in D. Orr and M. Soroos, ed., *The Global Predicament: Ecological Perspectives on World Order* (Chapel Hill: University of North Carolina, 1979).

[40] Robert Walker, *One World, Many Worlds: Struggles For a Just World Peace* (Boulder: Lynne Rienner, 1988), 60. See also the work of Charles Beitz and, especially, Richard Falk.

order, and many writers in this area were thinking in these terms about the environment long before mainstream political science began to take the question seriously. However, the normative imperative of globalism may be said to drive less radical analysts as well: Charles Pentland mentions the 'modernist' wing "among liberal institutionalists", thinkers such as Oran Young, which "accepts some idealist or globalist notions about the continuing development and the long-run transformational impact of interdependence", and many functionalists desired the eventual establishment of a politically united world.[41]

To many, then, prescriptive globalism is the call for deep changes in the social fabric of the world system. Attempts to deal with the environmental crises we face by cosmetically re-arranging the extant power structures are doomed to fail because it is those structures which are part of the cause of the problem in the first place. Suggestions that, for example, post-Cold War military alliances can help fight environmental degradation are dismissed as futile, if not obfuscatory. It is obvious that there is a close affinity between normative globalism and much of the environmentalist literature, even if the latter is often uninformed about world politics. But it is also true that much of the rhetoric we hear from politicians and statements or declarations released by international organizations assumes the cloak of globalism.

3. The Death of an Orthodoxy?

Literature linking security and ecology has increased in both volume and popularity in the last decade, and this is hardly surprising. Many scholars have been interested in providing an alternative conception of security, one based on threats to continued community survival which are not of a directly military nature. The role of natural resources in international conflict has been the object of sustained study in recent decades, and the Sprouts championed the concept of environmental context in world politics some time ago.[42] The field will continue to gain importance as scholars, politicians (including the new American vice-president), squadrons of strategic analysts, and international lawyers

[41] C. Pentland, *op. cit.*, 190.

[42] H. and M. Sprout, *Toward a Politics of the Planet Earth* (New York: van Nostrand Reinhold, 1971); and Dennis Pirages, who carries this tradition onward in "The Ecological Perspective and the Social Sciences", *International Studies Quarterly* 27 (1983), 243-255; and in his *Global Ecopolitics: The New Context for International Relations* (Massachusetts: Duxbury, 1978).

devote increased attention to related issues. This is, of course, part of a broader trend. Though it varies in intensity and perhaps will in durability, this "environmental enlightenment" has affected all the social sciences, including international security analysis.[43]

Predictably, mixed reactions -- reflective of assumptions and ideological predispositions -- greet this development. Environmentalists who have laboured on the front lines of radical activism for the lean years tend to view it with a cautionary and healthy cynicism. Beyond obligatory attention to the perennial philosophical questions about peace and war, the central operational *raison d'etre* of conventional security studies was the Cold War and its many secondary industries. Less apprehension is felt about including environmental concerns within the framework of what came to be known as peace studies, but some tension remains. Though the inclusion of ecological considerations in a general definition of security is welcomed, mainstream society has a way of co-opting social issues, even movements. If this new wave of thinking sacrifices the traditional and urgent call for global consciousness to the jealous deity of sovereignty, to the parochial world of national competition, little is gained. Much of the representative literature does just this: it adds environment to the national security paradigm and mixes.

Many strategic analysts, meanwhile, will probably respond to such a conceptual development with equally subdued enthusiasm, partly because they will rightly interpret some of the new literature as old geopolitics dressed in environmentalist garb. The real danger to them, however, may be that the sacred term security is itself in danger of losing its analytic, if not normative, significance. In a world of few constants, the need for a clearly defined security agenda is important in strategic analysis. It took an academic 'neorealist revolution' to legitimize the inclusion of economic factors in the traditional approach to national security.

---

43 Concern with environmental problems is reflected by a growing body of literature from a wide range of the social sciences. See the review essay by P. Lowe and W. Rudig, who complain that this literature has been "divided between too much uninformed and heavy-handed empiricism on the one hand and too much grand theorizing and pontificating on the other." "Review Article: Political Ecology and The Social Sciences -- The State of the Art", *British Journal of Political Science* 16:4 (1986), 513-550. See also S. Yearley, *The Green Case: A Sociology of Environmental Issues, Arguments and Politics* (London: HarperCollinsAcademic, 1991); S. Kamieniecki and E. Sanasarian, "Conducting Comparative Research on Environmental Policy", *Natural Resources Journal*, 30:2 (1990), 321-340; The "Special Curriculum Issue" of the *Environmental History Review* 16:1 (1992); M. McDougal and J. Schneider, "The Protection of the Environment and World Public Order: Some Recent Developments", *Mississippi Law Journal* 45 (1974), 1085-1124; and H. Brough, "Environmental Studies: Is it Academic?" *Worldwatch* 5:1 (1992), 26-33.

Adding this further complication not only distorts the billiard-ball imagery for which realism is so cursorily known, but it leads inexorably down the path to the often vague jargon characteristic of what we have earlier in this chapter defined as normative globalism.

There is a traditional friction between the two camps roughly sketched above. Environmental activists have often labelled military analysts as, at best, tools of an inherently dangerous and far-reaching military establishment. And in many places environmentalists are viewed as downright subversives.[44] This mutual hostility will no doubt characterize extreme reactions to the mainstreaming of environmental concerns. Many thinkers on both sides of the divide outlined above will habitually ignore those on the other regardless of what articles scholarly journals are printing, or what thematic conferences universities are holding. It is difficult to imagine a neoclassical strategist like Colin Gray reading with much serious interest the work of an ecological anarchist like Murray Bookchin, or vice-versa.

Relying less on caricature, it may be argued that recent concern with pressing environmental problems and the related causal connections with human conflict reflects not so much an ideological clash as concern shared across a wide spectrum of interested -- and, quite frankly, anxious -- minds. There is a perfectly logical, and sobering, reason why environmental concerns are now infiltrating even the most remote subject areas in contemporary western academia: the earth is, has been, and will undoubtedly continue to be in serious ecological trouble. Trends may reflect the momentum of fashion, but they are usually indicative of reality as well. The catalogue of environmental degradation around the globe has inspired a new round of what Jim MacNeil has termed the "greening of international relations,"[45] diplomatic activity which inevitably affects the study of international security. This is facilitated by the immediate international context: the demise of the Cold War, an uncertain but certainly nuclear future, the stagnation of the world economy, irrational dependencies upon certain natural resources, the unchanged plight of the underclasses of the world. Fears of Armageddon have subsided, but it is instead the gradual erosion of the planet that captures attention at this stage of human and biological history. This is partly due to the eclipse of

---

[44] A Brazilian War College Report, "Brazil 2,000", identified four problem communities in the Amazon region, in this order: environmentalists, indigenous peoples' organizations, smugglers, and drug traffickers. *Ploughshares Monitor*, December 1991, 6-7. Thanks to Fred Knelman for bringing this to my attention.

[45] "The Greening of International Relations", *International Journal* XLV (1989-90), 1-35.

the Cold War as the primary preoccupation in the study of world politics. Nuclear holocaust remains a dreaded possibility, but it no longer monopolizes the thinking of those concerned with planetary survival. Instead, we have become cognizant of the less spectacular, yet unrelenting, assault on earth from population growth, conventional warfare and economic activity. This has resulted in a challenge to the orthodoxy of security.

Barry Buzan, at the forefront of this challenge, makes sharp distinctions between strategic studies and security studies, and argues that the latter should focus on military, political, economic, societal, and environmental security. Buzan's initial thrust involved distinguishing between what he termed the national security and the international security strategies. Much of Buzan's popular book *People, States and Fear* is typical of literature dealing with this question: a frustrating circular discussion which ends in a predictable surrender to any grand aspirations it may have had.[46] Some mixture of national and international thinking in foreign policy is vital; too much of one is a bad thing, and too much of the other is highly unlikely. Buzan's utility lies in his reordering of the intellectual approaches to these questions. According to Buzan, the cognitive poverty of strategic analysis began after the Second World War, when power and security seemed diluted into interchangeable terms under the dominance of a realist paradigm.[47] Buzan calls for an expansion of academic security studies, autonomous from strategic analysis, since incorporating the former within the latter "would be like giving responsibility for designing a national transportation system to the makers of automobiles."[48]

This echoes other calls from mainstream literature. One of the more seminal publications, especially notable because it appeared in what is probably the most widely read scholarly journal in strategic studies, *International Security*, was written by Princeton's Richard Ullman and entitled, simply enough, "Redefining Security." Ullman offers a succinct ex-

---

[46] "The imperative of minimising vulnerabilities sits unhappily with the risks of international agreement, and the prospects of international agreement are weakened by the power-security dilemma effects of the national security strategy. Despite this problem, in the real world security policy must be, and indeed is, a mix, if only because the consequences of pursuing either strategy singlemindedly are so obviously disastrous." B. Buzan, *People, States, and Fear: The National Security Problem in International Relations* (Chapel Hill: University of North Carolina Press, 1983), 221. For a critique of Buzan's work see S. Dalby, *op. cit.*

[47] Barry Buzan, *People, States and Fear: An Agenda For International Security Studies in the Post-Cold War Era* (London: Harvester Wheatsheaf, 1991).

[48] *Ibid.*, 23.

position of what constitutes a threat to that often ambiguous priority, the national interest: "an action or sequence of events that (1) threatens drastically and over a relatively brief span of time to degrade the quality of life for the inhabitants of a state, or (2) threatens significantly to narrow the range of policy choices available to the government of a state or to private, nongovernmental entities within the state."[49] Surely the prospect of large parts of California falling into the Pacific, or hurricanes in Florida, or flooding in the midwest, or chronic fish shortages, constitute security threats equal to any contemporary military challenge? Though one might take issue with the somewhat superficial way he presents "Third World" problems, Ullman draws the link between the security of non-Americans and Americans in a mainstream journal accustomed to more parochial articles of faith (he also takes occasion to critique Reagan Administration policies in the early 1980s, especially with regard to its ideological reliance on the market mechanism for resource allocation and dependence on Persian Gulf oil). While it is commonly accepted that a strong centralized organization is necessary for military security, Ullman argues this acceptance should extend to other potential threats as well. The article is provocative and helped set the stage for further chipping away at the sacred policy position of military security, if not the canon of national security.

More American navel-gazing is supplied by Theodore Sorensen's "Rethinking National Security" which appeared in *Foreign Affairs*. Sorensen articulates the need, in the absence of an expansionist or ideological Soviet Union, for a new "national thesis". His practical proposal to this end is that the National Security Council be modified so as to include "fewer generals and kremlinologists and more economists and election analysts."[50] To the extent that American economic independence will

---

[49] R. Ullman, "Redefining Security", *International Security*, 8:1 (1983), 129-153; 133. Recent disclosures about the activities of the Federal Emergency Management Agency (FEMA) support Ullman's contention that not enough has been done to address non-military threats. FEMA has spent well over a billion dollars in the last decade "to make the government a moving target during a nuclear war", much to the anger of citizens who suffered in what was perceived as an inadequate response to Hurricane Andrew in Florida in August of 1992 and the San Francisco earthquake of 1989. According to a former Defence Department official with close ties to FEMA, "[p]eople still believed nuclear war was likely. People were not really thinking about hurricanes and earthquakes, they were thinking about war with the Soviet Union." "A Disaster Waiting to Happen", *The Globe and Mail*, Feb. 24, 1993 (Cox News Service).

[50] I suspect the kremlinologists have departed or altered their occupational titles. T. Sorensen, "Rethinking National Security", *Foreign Affairs* 69:3 (1990), 1-18. Sorensen's paper should not be perceived as radical in its orientation toward security. In fact, old notions of security are reified: "In a world that is still heavily armed, highly volatile, and increasingly complex, our

require energy security, Sorensen touches on the environmental debate. But much of this literature is really American introspection, "tinged", as Richard Falk said recently in a speech to Canadian lawyers, with a concern either about adjusting to a condition of relative decline, either along the cautionary lines advised by Paul Kennedy's influential book on the rise and fall of great powers, or how to go about regaining ascendancy in the world as proposed by such pundits as Henry Kissinger, Zbigniew Brzezinski, and ... Francis Fukuyama. Preoccupations in the United States, even when they purport to be concerned with issues of global policy, often turn out to be disguised inquiries into how the United States should go about playing a more successful role in the world system.[51]

This being said, it is obvious that the American role will be vital in forging the regulatory regimes needed to limit environmental degradation in the future. To this end, literature linking security with international environmental problems often calls for increased American assistance worldwide. Indeed, a standard approach has evolved, exemplified by Norman Myers, who looks at deforestation in the Philippines, water scarcity in the Middle East and Egypt, land degradation in El Salvador, and overpopulation in Mexico. After explaining what facet of American security is at risk, he proceeds to discuss these problems, avoiding their causal links with superpower security interests and American economic interests in general; and he ends by raising the spectre of hungry Mexicans (three million of them) heading north into America should conditions there lead to violent revolution.[52] Myers knows one sells products by appealing to insecurities.[53] Another example is the informative collection of essays edited by Janet Welsh Brown which calls for greater development assistance to deal with environmental problems which affect U.S. resource bases in the southern hemisphere.[54] At the very least, concerned analysts such as Jessica Tuchman Mathews have

---

instinctive obligations of national self-preservation and self-esteem require us to secure before all else the survival of our own nation's independence, institutions, and inhabitants" (page 3).

51 "The Capacity of International Law to Respond to the Environmental Challenge", Address to the Conference of the Canadian Council on International Law: Preserving the Global Environment; Ottawa, Oct. 19-21, 1989.

52 N. Myers, "Environment and Security", *Foreign Policy* 74 (Spring 1989), 23-41.

53 As does Nicholas Eberstadt, who writes of threats to western security related to rising southern, and declining northern, populations. See "Population Change and National Security", *Foreign Affairs* 70:3 (1991), 115-131.

54 J. Welsh Brown, ed., *In the U.S. Interest: Resources, Growth, and Security in the Developing World* (Boulder: Westview, 1990).

called for increased multilateral diplomacy, or American "leadership without primacy."[55]

The pro-environmental security position has perhaps been articulated the most clearly by Peter Gleick, who identifies four primary environmental threats to security, all relevant to resource studies. Resource acquisitions are strategic goals in themselves; they are often attacked as part of military strategy; resources can be utilized as military tools ("ecosabotage" is gaining currency as a concept, but this category has equally sinister implications in the use of food and water as weapons); and, finally, various disruptions to environmental services, such as water supply, are obvious threats to the well being of citizens.[56] It is thus necessary to view environmental threats within their proper context, as challenges to the national interest, and this invariably involves the participation of the organization whose explicated role in society is to protect that interest: the military establishment.

This is a serious problem for at least one analyst on the opposite end of the opinion spectrum, Daniel Deudney, who advises us to examine its temperament "before harnessing the old horse of national security to pull the heavy new environmental wagon"; he feels that viewing environmental security as a national interest problem risks sacrificing the world community approach to the problem.[57] Beyond this interesting level of analysis question, Deudney argues -- with less conviction -- that traditional security problems such as territorial control and resource acquisition should not really involve the environment as an issue-area. He maintains that we should not worry about resource wars. Today's trading system will provide ample resources, it is increasingly difficult to exploit foreign resources through territorial conquest, and we live in the age of substitution, where "industrial civilisation is increasingly capable of taking earth materials such as iron, aluminum, silicon and hydrocarbons (which are ubiquitous and plentiful) and fashioning them into virtually everything needed."[58] (He does not claim to be able to fashion sanitary

---

[55] J. Tuchman Mathews, "Redefining Security", *Foreign Affairs* 68:2 (1989), 175. This article contains a concise overview of contemporary global environmental problems.

[56] P. Gleick, "Environment and Security: The Clear Connections", *Bulletin of the Atomic Scientists* 47:3 (1991), 16-21.

[57] D. Deudney, "Environment and Security: Muddled Thinking", *Bulletin of the Atomic Scientists,* April 1991, 22-28, 24 and 25.

[58] Deudney, "The Mirage of Eco-War: The Weak Relationship Among Global Environmental Change, National Security and Interstate Violence", in I. Rowlands and M. Greene, *Global Environmental Change and International Relations* (London: Macmillan, 1992), 169-191. On the point of substitution, we might recall an assertion made two decades ago by Nazli Choucri:

drinking water, wood, or oil from these elements.) The mere fact that military establishments are the single largest polluters in many nations, a point we will discuss later, means that environmental considerations simply *must* be emphasized even in institutions based explicitly on a traditional, or national, concept of security.

In essence, Deudney's larger argument is about discourse. He asserts that treating environmental problems as security threats may subsume the former into mere categories of the latter. The world community must have priority. This is familiar phraseology: connections between global ecology and the need for world order have been made for some time, epitomized perhaps by Richard Falk's 1971 path-breaking *This Endangered Planet*, in which he asserted:

> A world of sovereign states is unable to cope with endangered-planet problems ... Such a system exhibits only a modest capacity for international cooperation and coordination. The distribution of power and authority, as well as the organization of human effort, is overwhelmingly guided by the selfish drives of nations.[59]

There is wide congruence in opinion on this last point. But the focus of the argument around the merits of state sovereignty as an international institution has rarely been as sharp. The point is worthy of repetition: "because the solutions to the problems of the environment must be global they will present an unprecedented challenge to concepts of national sovereignty."[60] Radical analysts look to transnational non-governmental organizations and movements as the agency of change, the

"[t]he problem here is one of organization and not of technological feasibility (although they are highly interdependent)." Too much faith in either strikes one as problematic. N. Choucri and J. Bennett, "Population, Resources, and Technology: Political Implications of the Environmental Crisis", in D. Kay and E. Skolnikoff, ed., *World Eco-Crisis: International Organizations in Response* (University of Wisconsin, 1972), 9-46. Or we might quote Edward Luttwak, with whom Deudney would probably have several disagreements: "It is one thing to develop a tolerable ersatz for tantalum or columbium needed by the pound ... quite another to replace oil by the million metric tonne." Luttwak actually argues that military intervention following the OPEC cartel price increase was justified. "Interaction and Access to Natural Resources", in H. Bull, ed., *Intervention in World Politics* (Oxford: Clavendon, 1984), 79-94. Of course, as Tuchman Mathews and others point out, it is renewable resources, and not the finite ones, that are presently threatened with severe scarcity.

[59] R. Falk, *This Endangered Planet: Prospects and Proposals for Human Survival* (New York: Random House, 1971), 37-38.

[60] David Newman, "The New Diplomatic Agenda: Are Governments Ready?" *International Affairs* 65:1 (1989), 29-42; 35.

key in the drive to -- in the carefully chosen word of one student -- make nation-states "biodegradable."[61] Robert Walker is one of the more elegant writers to question the analytic primacy of the state in our thinking about security. As a process wherein the "principle of hierarchical subordination gradually gave way to the principle of spatial exclusion", the historical development of the nation-state system produced the principle of state sovereignty, which, by offering both a spatial and a temporal resolution of the relationship between universality and particularity ... affirms a specific account of who we are -- citizens of particular states who have the potential to work toward universal standards of conduct by participating in statist political communities -- and denies the possibility of any other alternative.[62]

One such alternative, coincident with what has been labelled a normative globalist position, is a mixture of local identity and global consciousness. (True to Wolfer's warning, these alternative visions of security are also highly ambiguous.)

Thus the environmental security debate may be placed within a wider critique of a narrow definition of security. Africanist Timothy Shaw argues that security should be viewed in an unconventional manner outside the traditional Eurocentric framework since it "is inseparable from questions of the survival of states, classes, factions and property: the 'corporatist' nexus." This is certainly a valid point, since the violence which disrupts civilian life and creates such ecological and economic havoc in Africa takes the shape of overt class, ethnic, and racial warfare conducted by state oppression and guerrilla-led revolution.[63] Of course, this perspective opens a can of rather hungry worms, since it not only changes the focus of security questions from the state to groups or classes, which transcend national divisions themselves, but leads us inexorably to questions concerning state legitimacy, relations between state and society, and the stuff of governance itself.[64] This is not usually found in national

---

61 P. Wapner, *Making States Biodegradable: Ecological Activism and World Politics*, Phd. Dissertation, Princeton University, 1991; abstract on CD-ROM.

62 R.B.J. Walker, "Security, Sovereignty, and the Challenge of World Politics", *Alternatives* XV (1990), 3-27, 10 and 12.

63 T. Shaw, "Security Redefined: Unconventional Conflict in Africa", in S. Wright and J. Brownfoot, ed., *Africa in World Politics: Changing Perspectives* (London: Macmillan, 1987), 17-34, 18.

64 Thus security, as it is presently conceived, assumes a negative connotation for revolutionary thinkers, since it perpetuates unjust systems of exploitation. The national security rhetoric of the apartheid state of South Africa is a poignant example of how the term can mean radically different things to different segments of a population.

security analysts' theoretical inventories. But it is worth quoting Ken Booth, whose reflective critique of contemporary security studies ruffled a few feathers in the late 1970s:

> The prevailing Eurocentric definition of war results in our overlooking a considerable amount of human violence ... if we ignore the multitude of lesser conflicts which take place, we will underestimate the extent of social conflict throughout the world, and so tend to exaggerate the degree of order which exists.[65]

The work currently pursued by Thomas Homer-Dixon and others offers a sound corrective to this tendency. The Peace and Conflict Studies Program at University College, University of Toronto, has initiated a project on "Environmental Change and Acute Conflict." The project focuses on three issues which may lead to violence in southern nations: water scarcity and degradation, population displacement, and general economic decline.[66] This is a bold and progressive research agenda: if social factors produce ecological degradation which, in turn, induces violence amongst those affected, then it is incumbent upon specialists to study those social factors in great detail and this cannot be done within a conventional security studies framework. As challenging, and interesting, is dynamic historical work which takes ecological variables into account, such as the epic by Alfred Crosby which details the "biological expansion of Europe."[67]

There is room as well, as we have mentioned above, for normative considerations quite beyond the realm of everyday political analysis. Such is the fate of many studies of world order, such as the 1960's World Order Models Project, which dared to be explicitly normative despite the

---

[65] K. Booth, *Strategy and Ethnocentrism* (New York: Holmes and Meier, 1979), 145. For an entertaining reply to Booth and others like him, written with a touch of acerbity by a self-described "unregenerate realist", see C. Gray, "Through a Missile Tube Darkly: New Thinking' About Nuclear Strategy", *Political Studies* XLI:4 (1993), 661-671.

[66] For a succinct overview see T. Homer-Dixon, J. Boutwell, and G. Rathjens, "Environmental Change and Violent Conflict", *Scientific American,* February 1993, 38-45. The most cogent published explication of this thesis is written by Homer-Dixon, whose article is required reading for anyone interested in this issue. "On the Threshold: Environmental Changes and Causes of Acute Conflict", *International Security* 16:2 (1991), 76-116. See also O. Bennett, ed., *Greenwar: Environment and Conflict* (London: Panos Institute, 1991); Joyce Starr, "Water Wars", *Foreign Policy* 82 (Spring 1991), 17-36.

[67] A. Crosby, *Ecological Imperialism: The Biological Expansion of Europe, 900-1900* (Cambridge: Cambridge University Press, 1986). On "biohistory" in general, see S. Boyden, *BioHistory: The Interplay Between Human Society and Biosphere, Past and Present* (Paris: UNESCO, 1992).

prevailing bias against this sort of work at the time. Mel Gurtov provides a recent and readable example of such work, arguing that realism "contributes to perpetuating the disorderly world that justifies itself, in the manner of a doctor who keeps a patient on medication in order to ensure future visits."[68] Supporters of strategic studies (albeit a custom-modified version) freely admit that "the strategists' definition of security revolves around the interests of the state he serves, and thus operates to the detriment of a broader approach that emphasizes global peace and harmony."[69] Gurtov insists that bad values stand in the way of the more harmonious, equitable world perceived as possible by globalists: "Global Humanist" values include cooperation, spirituality, enoughness, community, and personal growth, while realist and what Gurtov calls "global-corporatist" values include such undesirables as competition, materialism, quantity, individualism, and the will to power. The implication is that we will find true security only by adopting the former set. Similarly, much of the recent *ecofeminist* literature emphasizes the role of values conducive to (and not contrary to) natural processes.[70] Since such authors would reject the coercive imposition of the desirable values (that would be self-defeating), they must either be spread by human agency, or, even more problematically, evolve as the defining part of some cognitive teleology. A historical materialist, meanwhile, would argue that values reflect the dominant means of production, and that human-nature relations have been defined by this exigency. The dialectical relationship evolves as we change nature and are thus changed by nature.[71] Our next chapter will broach these questions in greater detail.

Though they differ with respect to specifics, what most of the authors we have shortly introduced have in common is a tendency to ar-

---

[68] M. Gurtov, *Global Politics In the Human Interest* (Boulder: Lynne Rienner, 1988), 18.

[69] M. Fortmann and W. George, "Bridge Over Troubled Waters: Linking Strategic Studies and International Relations", in D. Haglund and M. Hawes, ed., *World Politics: Power, Interdependence and Dependence* (Toronto: Harcourt Brace Jovanovich, 1990), 456-481, 459.

[70] I. Diamond and G.F. Orenstein, eds., *Reweaving the World: The Emergence of Ecofeminism* (San Francisco: Sierra Club, 1990); V. Shiva, *Staying Alive: Women, Ecology, and Development* (London: Zed Books, 1988); V. Held, "Gender as an Influence on Cultural Norms Relating to War and the Environment", in A. Westing, ed., *Cultural Norms, War and the Environment* (Oxford: Oxford University Press, 1988), 44-52. Jacques Grinevald includes ecofeminism and ecopacifism within a general trend of *ecologism,* which he contrasts with militarism; these he asserts are the two modern contrasting civilizational trends. See his "The Greening of Europe", *Bulletin of Peace Proposals* 22:1 (1991), 41-47.

[71] For an interesting recent treatment see C. A. Aguirre Rojas, "Between Marx and Braudel: Making History, Knowing History", *Review (Fernand Braudel Center)*, 15:2 (1992), 175-219; and for classic text see F. Engels, *Dialectics of Nature,* Fifth Printing (Moscow: Progress, 1972).

gue that in some manner, mind-sets, world views, paradigms, or norms must change if a sustainable future awaits us. Ultimately, the question of agency remains pivotal, from an organizational, foreign policy and, even, personal perspective. In this context it is worth quoting at length the British scholar Fred Halliday, who offers one of the more challenging essays on redefining international politics:

> Exculpatory abstractions about history, system, structure and the like have little more validity than the invocation of the divine, or the astrological.We are all participants in the making of the future. The most apposite reply to Fukuyama, the new neo-Hegelian, is to recall the riposte of another eminent, if flawed, social scientist of the early nineteenth century, to the effect that while the philosophers have only interpreted the world, it was incumbent upon us to change it. This, the oldest item on the old agenda, may have much to recommend itself as the first item on the new.[72]

4. Conclusion

> "The achievement of even a fraction of the objectives sought by environmental analysts within a reasonable time period would be an achievement of monumental proportions."
>
> Dean Mann, 1991[73]

Security must be defined in at once a broader and simultaneously narrower sense. That is, we must attempt to achieve a global ontology in our outlook towards questions that lead us into the temptation of a parochial, nationalistic, perspective. Pontification aside, subsequent chapters in this book will make this much clearer, but we may certainly note that there is a wealth of literature which suggests the moment for such reconceptualization has come. At the same time, a definition of community, reasonable and meaningful, is essential if environmental security is to mean anything. We must transcend our usual proclivity to define our immediate security in terms of, say, how many missiles surround our enemies, and replace this with a more humane concern with how many trees surround our neighbourhoods. This being said, the missiles won't go away on their own.

---

[72] F. Halliday, "International Relations: Is There a New Agenda?" *Millennium: Journal of International Studies*, 20:1 (1991), 57-72, 71-72.

[73] D. Mann, "Environmental Learning in a Decentralized Political World", *Journal of International Affairs*, 44:2 (1991), 301-338, 306.

Defining environmental security on a national basis defies both the logic of environmental holism, and the extant ecological crisis. Yet it will be done because the systems of governance currently in operation are conducive to this approach. As such, one might argue, (with Deudney) that viewing the environmentalist cause (itself ill-defined and fractional) within the present security framework is an exercise in discursive co-option. Lothar Brock worries that, in the worst such case, "the concept of environmental security ... may be invoked to defend the status quo of the present world ecological order."[74] Such concerns are common, and remind one of Richard Ashley's insistence that "research and theory-building [should] avoid complicity in the very processes that contribute to the recurrence of violence and the pervasiveness of insecurity."[75] My intuition is that involving environmental problems in the study of security will, overall, bring us closer to the grammar of thought Ashley calls *rationality proper,* that "emancipatory logic" which "acknowledges the inescapable dependence of each person and social group on its environment."[76] It might even drive the linear logic of interstate politics circular.[77]

However, and viewed from any level of human organization, this is a conflictual world. The crises and opportunities afforded by a post Cold War earth in environmental decline open new vistas of conceptual inquiry and cooperative potential, but the nation-state system remains prone to confrontational politics and competitive economics. While geopolitics loses no allure in what is sure to be a conflictive global system, international organization acquires renewed significance as a global project and a field of relevant study. An enhanced view of environmental security can add to the agenda of this intricate network of political behavior, at both the intergovernmental and transnational levels. Ted Gurr is no doubt correct when he asserts that it is chimerical to assume that, even if conditions of scarcity worsen, wealthy states can be expected to accept radical proposals for change.[78] On the other hand, the pluralists, who to some extent rely on self-interest as an agent of cooperative change, will in

---

[74] L. Brock, "Peace Through Parks", *op. cit.*, 418.

[75] R. Ashley, *The Political Economy of War and Peace: The Sino-Soviet-American Triangle and the Modern Security Problematique* (New York: Nichols, 1980), 301 (original is in italics).

[76] *Ibid.*, 208 and 176.

[77] These geometric terms, well-chosen for their symbolism, are used by Gwyn Prins in "Politics and the Environment", *International Affairs*, 66:4 (1990), 711-730.

[78] Gurr refers specifically to Barbara Ward's "proposals for a 'global compact' and a 'world tax system' to close the gap between rich and poor nations." "On the Political Consequences of Scarcity and Economic Decline", *International Studies Quarterly*, 29:1 (1985), 51-75, 59. Gurr refers to Ward's *Progress for a Small Planet* (Harmondsworth: Penguin, 1979).

all likelihood appeal to increasingly anxious policy makers as the most reasonable.[79] The challenge is to incorporate environmental security into that conception of self-interest without encouraging the further parochialism of the human species, or perpetuating the processes of exploitation condemned by radical structuralists, and -- perhaps -- managing to encourage a globalist ethic in the process.

In the next chapter we will begin to specify what environmental security can actually mean, but at this stage we might emphasize the longitudinal dimension of the concept. There is little reason to discuss human efforts to preserve healthy ecosystems in the short term, and any definition of that current catchphrase, sustainable development, makes mention of the importance of passing on a livable world to our temporal neighbours. Given the immediacy of many threats, this appears difficult. Biodiversity issues involve the eternal extinction of species. Nuclear waste will be inherited by future generations for many centuries. Once significantly altered, a natural resource such as the Mekong River in Southeast Asia can never be completely the same. And yet in each case we will see that sufficient complexity exists to make even short-term solutions to the environmental hazards provided in these cases unlikely. Longer term solutions may already be beyond possibility, though it is argued here that intergovernmental and transnational organization will be imperative if they are to be found. This frustrating combination of urgency and complexity can be overwhelming. It could in effect alienate environmentalist and political theorist alike from their traditional willingness to explore new conceptualizations. The frustration is amplified in

[79] As Charles Pentland explains, "[f]unctionalists expect no miraculous overnight political conversions on the part of governments, even in Hobbesian conditions of extremity ... they rely on national self-interest to lure states into an array of enduring co-operative institutions." "Integration, Interdependence, and Institutions", *op. cit.*, 180. For a pragmatic but sweeping example see N. Schriver, "International Organization for Environmental Security", *Bulletin of Peace Proposals*, 20:2 (1989), 115-122.

this age of high technology, and what David Harvey has termed *time-space compression,* "characterized by speed-up in the pace of life, while so overcoming spatial barriers that the world sometimes seems to collapse inwards upon us."[80] Therein lies, if the reader will tolerate the literary employment of such an egregiously fashionable term, the angst of the matter.

80 Time-space compression takes place as "space appears to shrink to a 'global village' of telecommunications and 'spaceship earth' of economic and ecological interdependencies -- to use just two familiar and everyday images -- and as time horizons shorten to the point where the present is all there is ..." D. Harvey, *The Condition of Post-Modernity: An Enquiry Into the Origins of Cultural Change* (Cambridge, MA: Blackwell, 1989), 240.

CHAPTER THREE

GLOBAL ENVIRONMENTAL SECURITY AND INTERNATIONAL ORGANIZATION

"Today ... people ... feel policymakers may be unduly concerned with the 'interest of all of mankind.' They see them sacrificing the less inclusive national community to the wider but in their opinion chimeric world community. The issue, then, is not one of transcending narrow group selfishness, as it was at the time [1930s] of [Charles] Beard's discussion, but rather one of according more exclusive devotion to the narrower cause of the national self."
Arnold Wolfers, 1952[1]

"Biospheric egalitarianism and the principles of ecological interconnection help us to realize that no large-scale impacts on ecosystems will be without their effects on human life."
A.R. Drengson, 1980[2]

1. INTRODUCTION

We have seen that the unprecedented attention given to the ecological imperatives of our time contributes to a growing tendency in published literature to strive for a contemporarily germane redefinition of the very concept of security. Ultimately, the trek towards this new conception will prove a familiarly divisive one as those whose primary loyalty is to the vague notion of "national interest" and those who insist on an even

[1] A. Wolfers, "National Security as an Ambiguous Symbol", *Political Science Quarterly*, LXVII:4 (1952), 481-502; 482.

[2] A.R. Drengson, "Shifting Paradigms: From the Technocratic to the Person-Planetary", *Environmental Ethics*, 2 (1980), 221-240.

less clear "globalist" definition emerge as vocal or prolific proponents. However, we may rest assured that this divergence, linked by a common concern with yet a third level of security, the personal, can enhance the dynamic discourses of international politics, foreign policy, transnational affairs, and ecology.

The redefinition of security may be seen as a vital first step towards a normative transition which has often been portrayed (perhaps quixotically) by environmentalists as either inevitable or axiomatic. Though this resembles a slow and painful revaluation rather than some remarkable long-awaited paradigm shift, it is discernible. The United Nations Conference on Environment and Development (UNCED) was of course a disappointment to those who thought the world would be saved in two weeks in Rio de Janeiro in 1992, but it also symbolized the near universal acceptance of the importance of the ongoing discourse of "sustainable development", or what has been termed *ecopolitics.* Compared to the dismal Cancun Mexico Summit of 1981 (which attracted only 22 participant states), UNCED marks a new, and much more serious, North-South dialogue.

Global environmental security (GES) is offered here as a unifying concept in the sense that it bonds the natural and social sciences in common cause, and promotes that rarest of commodities, integrated policy thinking about the future. However, its ability to unite should not be exaggerated. The usual ontological (and even ideological) divisions which characterize the various relevant disciplines will invariably surface. This should be viewed as positive, partly because a staunch dislike of orthodoxy is intellectually healthy but also because the related discourse will have a better attended future if it is fuelled by debate. This chapter will offer an exploratory definition of global environmental security, and discuss five of its complex dimensions. We will then discuss the relationship between GES and the discipline to which this work is intended as a contribution, the study of international organization, and offer some brief examples of GES problems before advancing to our more specific case studies.

A theme quite common in the study of international relations emerged in the previous chapter: the contradiction between the increasing urgency of the situation and the slow pace at which its articulation in organized political behavior can take place. There is a frustrating disharmony between what needs doing and what, given the divergent policy directions pursued by different political actors and the weight of structural constraints imposed upon them, can be done in the near future. Substantial policy initiatives, multilateral and transnational, have surfaced in the international arena; and though their motivational source remains a complex and contentious topic, we can understand manifestations of international environmental concern as indicators of normative change. But it has already become a tired axiom to say that this is all too little, too late. That it is too little, given the magnitude of problems such as desertification and deforestation, endangered species, or ocean pollution, is self evident. But if it is too late, we move from the realm of dissatisfaction to despair. This is not a plea for groundless optimism (on the contrary, for that is as dangerous a thing). But we must be able to accept the incremental nature of world transition, and to work within this temporal framework. In this way, rather than implicitly rejecting calls for truly radical change, we accept their long-term possibility.

2. GLOBAL ENVIRONMENTAL SECURITY (GES)

As we saw in the preceding chapter, the academic challenge to the traditional definition of security in international relations has become a mini-industry. We have also seen that there is an intrinsic debate within this field as to whether environmental problems should be viewed from a security framework. My answer is yes, they should: but we need to rework our approach to security. Few terms are used with similar promiscuity in political literature. At the heart of the present enterprise is the guiding assumption that the process of examining non-traditional threats (or, perhaps, examining traditional threats in non-traditional ways) and international responses will provoke this conceptual alteration. This is a typically ethnocentric -- indeed, even egocentric -- proposition, for other inhabitants of the earth no doubt have viewed security as differently as they view other essential facets of life. There are rough universals here. For example, personal and familial security is in most cases going to take precedence over others. But we can say with some certainty that national security has usually been defined in relation to threats to the sovereignty or national interests -- whatever those terms are assumed to mean -- of states. A military definition of security, which leads to the security dilemma discussed by John Herz, Robert Jervis, and others, has prevailed.[3] This dilemma is a real one, and this study is not denying its existence. To borrow from the analysis by Thucydides: provided its leaders

[3] See J. Herz, "Idealist Internationalism and The Security Dilemma", *World Politics*, 2 (1950), 157-180; and R. Jervis, "Cooperation Under the Security Dilemma", *World Politics*, 30 (1978), 167-212.

have been trained to think in such terms, Sparta's officials will always view an increase in Athenian capabilities with grave suspicion. What has changed is that we now know that we face collective threats, and ones which can only be met with sustained cooperation. The dilemma has grown considerably universal in this sense.

It is not difficult to explain why environmental security is vital. The key to the survival of any species, providing it is not being hunted directly by an overwhelming force, is habitat. While the exact requirements for environmental health are of course debatable, we can draw some general conclusions based on our knowledge of the biosphere. This is elementary, and yet what we might term its praxiological corollary -- that, in order to provide for our survival, we must protect our habitat -- is among the most divisive and confusing of aphorisms, diminished in relevance by the quest of western culture to separate itself from nature. In the age of the nation-state as the organizing principle of humanity, habitat has been replaced by territory, which in turn is defined by the existence of an outside world complete with real or potential enemies. The need for environmental security, and its correspondence to some image of community, is less controversial to those peoples currently threatened with physical and cultural extinction by anthropogenic habitat destruction, such as the Yanomamo Indians of Amazonia; or the Saami people in Arctic Fenno-Scandia, who lost much of their livelihood because the Chernobyl meltdown[4] raised the radioactivity level of their reindeer to unacceptable levels; or the Pitjantjatjaran Aborigines in South Australia, displaced due to British nuclear tests conducted from 1953-1963.[5] Of course, this dreary list goes on and on, ending in the ultimate threat to sustainable life on earth presented by nuclear weapons, ozone layer depletion, and global climate change. Thus the irony of present developments: in order to expand our definition of security we had to threaten its very core, habitat, and at its most encompassing level, that of the biosphere. In this light, the terms sustainability and security should be interchangeable.

The present author is compelled to explicate what, exactly, he ambiguously enough refers to as an "environmentalist perspective".[6]

---

[4] I.G. Simmons, *Changing the Face of the Earth: Culture, Environment, History* (Oxford: Basil Blackwell, 1989), 270.

[5] E. Nelson, "Britain's Aboriginal Sin", *Bulletin of the Atomic Scientists*, 48:6 (1992), 8-9.

[6] Overviews of the many different strands of modern environmentalist thought may be found in K. Coates, ed., *Socialism and the Environment* (Nottingham, 1972); W. Devall and G. Sessions, *Deep Ecology: Living as if Nature Mattered* (Salt Lake City: Peregrine, 1985); J. McCormick, *Reclaiming Paradise: The Global Environmental Movement* (Indianapolis, 1989); T. O'Riordan, *Environmentalism* (London, 1976); T. Roszak, *Where the Wasteland Ends: Politics and*

Of course, there are as many understandings of this term as there are of world politics, and one can but give a brief synopsis of a personal position here. It is informed by a normative concern for social justice and a rejection of what Richard Falk has termed the "technocratic outlook [which] is a continuation of that grave mistake of supposing that human society could successfully separate itself from nature"[7] as well as the myth of unlimited economic growth and the belief that industrialization is the only development option open to human communities. As such, any environmentalist perspective will be fundamentally concerned with the ecological effects of human activity and seek to mitigate potential or actual environmental harm; though there is considerable disagreement as to what, exactly, constitutes this harm and what should be done about it. It should be mentioned that this book is not based on an exceptionally militant environmentalism, though it is one which asserts the contemporary priority of ecological considerations and recognizes the inherent value of wildlife. Schemes which envision radically coercive environmentalism present serious problems; it is difficult for example not to agree with John Young when he writes that some "[d]eep ecologists are probably wasting their time if they expect majorities to surrender their limited freedom for some 'eco-fascist' future which attempts to save species at the expense of the established social and political values of democratic society."[8] Worse, authoritarian solutions would no doubt necessitate the establishment of a highly centralized enforcement mechanism, and much environmentalist thought has been predicated upon the political agenda of decentralization. This is why reflexive calls for world government are not only riddled with practical difficulties, but are open to conceptual challenge from within the very camp that often makes them.

---

transcendence in Postindustrial Society (Garden City: Doubleday, 1972); J. Young, *Post Environmentalism* (London, 1990); R. Sack, "The Realm of Meaning: The Inadequacy of Human-Nature Theory and the View of Mass Consumption", in B. Turner, ed., *The Earth As Transformed By Human Action: Global and Regional Changes in the Biosphere Over the Past 300 Years* (Cambridge University Press, 1990), 659-671; D. VanDeVeer and C. Pierce, ed., *People, Penguins, and Plastic Trees: Basic Issues in Environmental Ethics* (Belmont: Wadsworth, 1986); and many other relevant works.

[7] R. Falk, *The Promise of World Order: Essays in Normative International Relations* (Philadelphia: Temple University Press, 1987), 8.

[8] J. Young, *op. cit.*, 147. We should note that Young is not using the term "deep ecologists", most of whom are anything but "eco-fascists", in any specific capacity. He is identifying a possible, perhaps logical, progression in thought.

We may at this point offer a generic definition of environmental security.

At minimum, environmental security must imply that human and non-human rights to an environment free from destruction exist. Destruction may be defined in terms of sustainability: taking from nature does not constitute a threat in itself, and is of course essential for human survival. But taking from, or in some manner contributing to the degradation of or radically altering nature at irreplaceable levels is the prime threat to GES. Sustainable development, which should be seen as a universal concept that doesn't just apply to the "underdeveloped" hemisphere, implies that what we do to the environment today will not prohibit future generations from enjoying life.[9] This may be taken at a variety of levels (local, regional, national, continental, global); and it is meant to apply to a broad range of social and ecological groupings, such as families, small communities, ecosystems, wildlife refuges, cities, and non-state nations. It could even incorporate ethnicity and race, as ethnocide and ecocide are usually partners in destruction.

In short, there is no need to be bound by a concise level of analysis in our definition, geographic or otherwise, though common sense suggests that in the long run the planetary level is of utmost importance to all others. Nor is there any form of the precise measurement of such a thing; it is a subjective concept, implicitly unrestrained by the dictates of positivist analysis. We can, however, identify circumstances where there has been an absolute loss of environmental security, such as the poisonous night of the pesticide-gas leak (December 2-3, 1984) in Bhopal, India; the necessary closing of the cod fisheries off the coast of Canada in 1992-1993; the exodus of Ukrainian citizens from the Chernobyl area after the accident there; and so many other instances when people or other mammals have been deprived due to human interference with nature. We can also identify certain threats to GES, such as the arms race, with its attendant radioactive contamination in both superpower states; the continuation of hazardous marine oil transportation over large distances; or rising populations in the South and consumption patterns in the North. The time factor is relevant also, since more immediate threats, like the France-Japan

[9] One criticism of the UNCED process is that, by explicitly linking environment and development, ecological crises are seen as caused by poverty. It is in fact relative affluence that has caused much direct pollution and extraction of resources from the biosphere. This being said, it is vital to understand the links between environment and development options, and North-South transfers in this respect are obviously necessary.

overseas plutonium shipments, differ in their implications from long-term trends such as global warming.

This is indeed a slippery slope, since most forms of major environmental alteration by humans -- and some of this is of course essential for human survival -- could be classified as threats to GES. Our subsequent discussion will, it is hoped, address this question. We should be perfectly willing to accept the notion that GES, like any other concept in social science, is a relative thing, and will be defined by social actors in ways reflective of their own ideological predispositions. This includes what we may term "human-nature ideology" as well. How actors conceive of their ethical obligations towards the environment is a fundamental component of international political economy. This being said, there are four primary dimensions of GES that the present author feels are essential, and we will discuss them each in brief turn. They include the mitigation of threats to the basic health needs of ecosystems; political decision-making by state and non-state actors which facilitates risk aversion; human rights and equality of access to environmental security in secular and temporal terms; and the old international relations theory chestnut itself, the question of sovereignty.

I. Biospheric Threats

Barry Buzan defines environmental security as a concern with "the maintenance of the local and planetary biosphere as the essential support system on which all other human enterprises depend."[10] This leaves rather a great deal to interpretation, but at least makes clear a central tenet of ecology, that of physical interdependence. Environmental security for one country only is a patently absurd idea, and the nexus between global and local concerns is the biosphere itself, the part of the earth that contains living organisms. Recognition of the existence of such a colossal structure led to new thinking about ecology, and even the so-called Gaia hypothesis, which suggested that the entire ecosphere is analogous to a single living organism.[11] As Lynton Caldwell writes in his extensive survey of international environmental policy:

---

[10] B. Buzan, "New Patterns of Global Security in the Twenty-First Century", *International Affairs*, 67:3 (1991), 431-451, 433.

[11] The classic text is J.E. Lovelock's *Gaia: A New Look at Life on Earth* (New York: Oxford University Press, 1979). Gaia was the Greek Goddess of the Earth. For those interested in exploring the bond between humans and the biosphere of which they are part, I highly

> Once the biosphere concept was accepted, human impact upon the environment took on a new significance. Whereas an earlier generation had worried about the health effects of coal smoke in the air over London and Pittsburgh, informed people in the 1960's became concerned about burning coal for energy and the resulting carbon dioxide balance in the earth's atmosphere.[12]

Security would be a meaningless concept without its opposite, insecurity, which is usually defined by the presence of threat, regardless of the object to be protected. A state-centric perspective in international relations that assumes the existence of a security dilemma among nations obviously places great emphasis on the role of national governments in identifying, and curtailing, threats to physically identifiable boundaries (territorial integrity) and much less clearly identifiable social constructs, such as national sovereignty. As the growing debates over the contemporary meaning of the term indicate, security -- national or global -- is a subjective concept, and threats are often suggested if not predetermined by a dominant paradigm.

Similarly, there will always be tension between conventional and more holistic approaches to ecology, and this is of epistemological importance with regard to the vital need for strong links between the natural and social sciences.[13] But the ecological crises exists primarily because of the empirical identification of environmental degradation on a planetary scale. Though our instruments remain imprecise and subject to political manipulation, depleting fish stocks, agricultural yields, and even holes in the stratospheric ozone layer can be measured. Relatedly, we can quantify the unemployment resulting from Canada's east coast moratorium on cod fisheries, the movement of "environmental refugees" in soil-eroded Africa, and the increase in Australian skin cancer patients. Viewed globally, all of these developments must be seen as threats to biospheric integrity; the political question is one of assigned priority and issue longevity. There is, then, what we may with perhaps some trepidation term

---

recommend E.O. Wilson's *Biophilia: The Human Bond With Other Species* (London: Cambridge University Press, 1984).

[12] L. Caldwell, *International Environmental Policy: Emergence and Dimensions*, Second Edition (Duke: London, 1990), 29.

[13] For a concise discussion of the role of positivist and holistic science in environmental decision-making and an example (the Spruce Budworm spraying controversy in New Brunswick), see Alan Miller, "The Role of Analytical Science in Natural Resource Decision-Making", *Environmental Management*, 17:5 (1993), 563-574.

a *scientific grounding* for GES: an identification of human-induced challenges to the sustainability of ecosystems.

This book is not intended as an exercise in environmental science, but we may briefly present what S. Boyden has declared the "basic health needs of the ecosystems of the biosphere" (see Figure 3.1). The biosphere may be conceived as the ultimate earthly morphic unit, composed of a series of interrelated ecosystems.[14] Ecosystems are systems in which "there is an interdependence upon and interaction between living organisms and their immediate physical, chemical and biological environment. Ecosystems operate on a wide variety of scales, ranging from tropical rainforests to small rock pools and in each nutrients and matter move continuously between the various components, often in well-defined cyclical pathways." The biosphere serves to provide habitat for living organisms to live and evolve, and it also "forms a self-regenerating system in which energy is provided by the sun and the materials essential for life are recycled from within the system."[15] Readers with a background in world politics theory will probably note a resonance with the systems approaches that surfaced in the discipline in the 1950s and 1960s, and which offered an interesting ontological challenge at the time.[16]

We see, then, that the specific health needs of ecosystems are contingent on the attainment of biospheric, or global environmental security; and, of course and always, vice-versa. While we will look at the larger global threats within this chapter, our case studies will center on what have been more limited threats to ecosystems: loss of biodiversity through species reduction; possible spread of radioactive material; environmental alteration during development projects; and population displacement related to ecological deterioration. All of these instances present immediate threats to biospheric integrity, and they have all involved significant activity at the international level, with varying results in terms of scientific consensus and normative change.

---

[14] A morphic unit is defined by Rupert Sheldrake as a "unit of form or organization, such as an atom, molecule, crystal, cell, plant, animal, pattern of instinctive behaviour, social group, element of culture, ecosystem, planet, planetary system, or galaxy. [They] are organized in nested hierarchies of units within units" of infinite smallness, and largeness. Every thing is a morphic unit; everything is composed of morphic units. This is not as useless a formula as it sounds, since the intellectual activity it can engender, identifying and analyzing the links between various units, is the essence of science, social and natural. Sheldrake, *The Presence of the Past: Morphic Resonance and the Habits of Nature* (New York: Vintage, 1989), 571.

[15] Both the preceding quotes are from the very helpful Collins Reference Dictionary, *Environmental Science* (London: Collins, 1990), pages 140 and 55.

[16] See esp. E. Laszlo, *The Systems View of the World* (New York: Braziller, 1972).

Figure 3.1: Basic Health Needs of the Ecosystems of the Biosphere
1. A rate of soil loss no greater than the rate of soil formation;
2. Intact nutrient cycles maintaining the nutrient properties of the soil over long periods of time;
3. The absence of polluting gases or particles in the atmosphere which interfere with living processes or significantly modify the climate;
4. The maintenance of an intact ozone layer in the stratosphere protecting the Earth's surface from ultraviolet radiation from the sun;
5. The absence, in oceans, lakes, rivers, and streams and in the soil, of concentrations of chemical compounds likely to be harmful to living organisms;*
6. The absence of levels of ionising radiation that can interfere with the normal processes of life and of bioreproduction;*
7. The maintenance of biological diversity.*

Source: S. Boyden, *BioHistory: The Interplay Between Human Society and the Biosphere, Past and Present* (Paris: UNESCO, 1992); page 26. I have put asterisks beside the health needs that the case studies will explicitly deal with in this book.

II. Decision-Making and Risk Aversion

At the immediately political level, GES is about focused or non-focused environmental risk-related decision-making in an international or transnational forum. This includes the plethora of foreign policy decisions, and national policy decisions, that effect -- positively or negatively -- the ecosystem health needs outlined above, as well as the participatory decisions of the multitude of other political actors involved in these matters. Focused environmental consideration has, of course, increased in scope and depth in recent years; and this has occurred, as our case studies of the IWC, IAEA, and MDC will to varying degrees demonstrate, within international organizations themselves. But it is as important to include, insofar as they can be identified, non-focused, or incidental, decisions taken which affect the environment. The Iraqi decision to invade Kuwait would prove disastrous for the environment, but this consideration clearly had little impact, if it even was one. Multinational mining firms are investing heavily in Argentina not because of that countries' commitment to environmental protection, but at least in part because of its related laxity. Most decisions that effect the environment do so as an indirect consequence; this includes profit-maximization in the global marketplace, military-strategic policies that are designed to protect or destroy territorial spaces, structural adjustment programmes intended to achieve balance-of-payments harmony, and countless others. At the international level, where there is no central authority to make and enforce environmental decision-making, we are bound to examine resource regimes and other forms of diplomatic cooperation, industrial and agricultural production and trade patterns, and non-state activism. We will discuss this in more detail in the last section of this chapter. However, we must be aware of the relative position of environmental concerns in terms of international priorities.

Environmental decision-making is based on perceptions of risks (threats to the biosphere), and the protection of interests. The two coincide in the political arena. For example, one argument against nuclear power, in addition to the ecological threats mentioned above, is that it poses additional risks, such as weapons proliferation and terrorist susceptibility.[17] Heavy reliance on imported oil, on the other hand, risks economic vulnerability to political developments in the Middle East or, in Japan's case, the stability of the hotly disputed Southeast Asian sea lanes. It also risks catastrophic oil spills, such as the infamous Exxon *Valdez* off the coast of Alaska. Large scale hydroelectric development, as Hydro Quebec of Canada (which miscalculated high demand from New York State) discovered, often depends on foreign energy markets which are being reduced due to conservationist measures elsewhere. And the environment may be victim to external dependencies: Laos' greatest source of foreign exchange is in hydro energy exports to Thailand.

The environmental impact of the culmination of related decisions is not always visible, as the greenhouse effect suggests, but we can iden-

---

[17] Though the terrorist argument has come and gone over the years, it is now picking up renewed strength in the post Cold War era. This may be related to the bombing of the World Trade Center, and increased anxiety over terrorist activity in such areas as the Middle East and Northern Ireland. See M. Wald, "Senators Want Atomic Plants Protected Against Car Bombs", *New York Times*, March 20, 1994, 14. Senator Max Baucus, chairman of the Committee on Environment and Public Works, asserts that "Our nuclear power plants should be at least as well protected as our politicians."

tify key actors, from individual consumers to multinational executives and politicians, involved in the process.

Emphasizing the decision-making element may draw the ire of structuralists of all stripes. But there is no ulterior agenda designed to deny the impact of structure here. This recalls our previous discussion of the dual nature of regimes, which serve to not only proscribe behavior but also to legitimize it. Such institutions, which operate in a world characterized by the deeper structures of history, become structural themselves and form part of the context in which further decisions are made. Structures exist primarily as patterns, not laws, though beliefs in structural laws, physical and social, influence the patterns. As Ernst Haas writes, those he labels (rather sweepingly) as *eco-reformers* want structures to be both laws and patterns:

> If it is the former the world system has a life of its own, which limits the choice left to the actor; if it is the latter the system is of the more open variety, in which causation flows from the actors to the constraints. Most eco-reformers prefer to have it both ways. They obviously want the actors to master the problematique, an achievement that depends on choice and will; but they also cherish the legitimacy that comes from being able to argue that 'science, our values, and our analysis are all congruent.' The imputation given to structures, therefore, partakes of both meanings.[18]

Differing perspectives on world politics, even those roughly outlined in the first chapter of this book, will have different views of what is known as the Agent-Structure Debate. It is simply suggested here that a rejection of the determinism inherent in adherence to one side or the other is in order. It is also important to recognize the importance of power, both in terms of its diffuse nature in today's world and its retained concentration in the hands and minds of the leaders of nation-states.[19] Even

---

[18] E. Haas, "Words Can Hurt You; Or, Who Said What to Whom About Regimes", in S. Krasner, ed., *International Regimes* (Ithaca: Cornell, 1983), 23-60, 41. Despite its arbitrary -- and often quite confusing -- taxonomy of thinking about regimes, this remains the most profound reading on the question.

[19] As notes Christopher Stone, "[e]very individual has the opportunity to be either cruel or kind to animals in his or her personal conduct, and is thus disposed to absorb norms on the subject. But who has the power and opportunity to defoliate jungles or to precipitate tsunamis (seismic sea waves)?" "The Law as a Force in Shaping Cultural Norms Relating to

the most committed globalists cannot wish away the incredible technological power, manifested in an ability to both create and destroy, wielded by military decision-makers and industrial executives alike. At the same time, these people operate in the social context of political structures that limit their actions and, in the case for example of the global marketplace, often determine them. Likewise, NGOs can have a significant impact on decision-making, but their efficacy is obviously limited by their relative power. And so this author shamelessly accepts an eclectic approach to the general question of causation in world politics.[20]

Without venturing into the hazardous and quite often helplessly esoteric waters of decision-making theory, we must of course acknowledge the chaotic environment in which political decisions are made, and not only by governmental actors but the plethora of participants in trends in GES. We all possess imperfect information, and we are all somewhat less than perfectly rational (therefore, of course, no one can be sure perfect information or rationality exists). The *precautionary principle* has received much attention as of late within environmentalism's broadening discourse. If decision-makers are not very sure that a human activity will result in little or no environmental damage, they should err on the side of caution. Principle Fifteen of the Rio Declaration alludes to this:

> the precautionary approach shall be widely applied by States according to their capabilities. Where there are threats of serious or irreversible damage, lack of full scientific certainty shall not be used as a reason for postponing cost-effective measures to prevent environmental degradation.[21]

Environmental decision-making, whether it be foreign policy, domestic policy, or some complex mixture of the two (the most likely

---

War and the Environment", in A. Westing, ed., *Cultural Norms, War and the Environment* (Oxford: Oxford University Press, 1988), 64-82, 70-71.

[20] In this connection I quote Seyom Brown, who accepts a similar verdict: "This is the explanation I come to, not from a wimpy eclecticism, but rather out of a philosophical/analytical conviction ... that change or stability can be generated at any level (from micro to macro) in the world polity ... If this be heresy to those who worship at the god of theoretical parsimony, so be it." S. Brown, "Explaining the Transformation of World Politics", *International Journal*, XLVI:2 (1991), 207-219, 217-218. For another recent and interesting discussion which advocates a "competing authority patterns" framework, see R. Mansbach, "The World Turned Upside Down", *The Journal of East Asian Affairs*, 7:2 (1993), 451-497.

[21] See also B. Wynne, "Uncertainty and Environmental Learning: Reconceiving Science and Policy in the Preventive Paradigm", *Global Environmental Change*, 2:2 (1992), 111-127.

scenario), will always come down to a question of the relative benefits, and to whom the benefits will accrue, of any decision made. And the option of making no real decision at all is always attractive to politicians and bureaucrats alike. All this occurs in a world shifted by constant struggle at the social level of human interaction. Beyond this, there is the question of how international organizations, for example, can come to positions on issues in the first place. This is a fascinatingly dynamic decision-making environment, reflective of both those that influence the organization and, to a limited extent, the more-or-less autonomous body incorporated in the institution itself. The future will be about hard choices, and the spaces in which those choices are framed will have great influence. No one is underestimating the difficulty involved. Consider the following rumination of Mostafa Tolba, previous Director-General of the UNEP, when asked about future concerns: "we have to control acid rain by controlling the emissions of sulphur, and sulphur dioxide. But particles of sulphur cut down on global warming ... so there may have to be a trade off [and] that is one issue that is going to harass the environmental agenda in the next years."[22] It is a question of international governance, and what is needed, and what may be developing in a very slow but discernible way, is an admittedly fragile ability through international organization to change existing regimes or create new ones that articulate changing conceptions of GES risks.[23]

III. The Normative Context: GES as a Human Right

With this question we move irrevocably beyond the analytic and into the normative realm. The right to existence is contingent on a conducive, inhabitable, environment, access to which is as clear an indicator of social equality as any. This is a contentious issue even at the level of the global commons, where, a normative globalist might argue, it should be most obvious.[24] In this regard the North-South divide becomes central,

---

[22] Quoted in *Our Planet* (UNEP), 4:6 (1992), 8-11.

[23] For a related idea, see Shiping Zhang,"Environmental Rights and the Global Common Risk Community", paper presented to the Annual Conference of the Canadian Sociology and Anthropology Association, Ottawa, 4-7 June 1993. For a recent discussion and case study dealing with international compliance with such regimes, see R.B. Mitchell, Intentional Oil Pollution at Sea: Environmental Policy and Treaty Compliance (Cambridge, MIT Press, 1994).

[24] P.M. Wijkman, "Managing the Global Commons", *International Organization*, 36:3 (1982), 511-536.

and will continue to be in any intergovernmental negotiations.[25] Principle Seven of the Rio Declaration makes note of the "different contributions to global degradation" and resulting "common but differentiated responsibilities."[26] One may go beyond a simple centre-periphery world model, however, to suggest that environmental security will differ according to class, regardless of location. The process of Westernization, then, becomes an example of shifting locations and meanings of environmental security. Alex Shoumatoff reflects upon this during his voyages in the Brazilian Amazon:

> This wasn't really as much an environmental battle, I decided, as a class struggle and a war of conquest. The tappers were being displaced by the ranchers, the agents of progress, the vanguard of a "superior" culture, just as the tappers had displaced the Indians, just as we had done in the American West, as has happened again and again all over the world from time immemorial. The interesting thing was that now, in the late twentieth century, you couldn't get away with it ... The eyes of your fellow global villagers were upon you...[27]

Access must be seen as a human rights question which cuts across all levels, from the personal to the global. This is of course easier said than done, and the picture is clouded considerably with the introduction of the dimension of time. If sustainable development means anything, it means that future generations will have as much environmental security as the present. This is complicated by present inequalities, which suggest that progress must be about improvement and not simple conservation, for most of the human race. Perhaps we are so caught up in a race driven by what has been termed *diachronic competition* ("a relationship whereby

---

[25] R. E. Benedick concludes: "As a consequence of the ozone issue, the richer nations for the first time acknowledged a responsibility to help developing countries to implement needed environmental policies without sacrificing aspirations for improved standards of living." *Ozone Diplomacy: New Directions in Safeguarding the Planet* (Cambridge: Harvard, 1991), 207. See also R.R. White, "Environmental Management and National Sovereignty: Some Issues From Senegal", *International Journal*, XLV (Winter, 1990), 106-137.

[26] More precisely: "The developed countries acknowledge the responsibility that they bear in the international pursuit of sustainable development in view of the pressures their societies place on the global environment and of the technologies and financial resources they command."

[27] A. Shoumatoff, *The World Is Burning* (Boston: Little, Brown and Company, 1990), 191.

contemporary well-being is achieved at the expense of our descendants")[28] that viewing the well-being of future humans as a policy priority is beyond the abilities of our decision-making mechanisms. Nonetheless this remains a logical goal, both of organizational objectives and the articulation in international law of such principles.[29] And, as Allen Springer writes, "'aspirational-rhetorical' equities provide an attitudinal context in which states approach environmental issues."[30] If this is so, the predictably disappointing Earth Summit should be seen in a much more productive light. Principle Six of the Rio Declaration refers to the "special priority" that should be given to the needs of the "most environmentally vulnerable nations"; as importantly, Principle Three declares that "The right to development must be fulfilled so as to equitably meet the developmental and environmental needs of present and future generations."

The Declaration is less enthusiastic about establishing a firm foundation for the evolution of a conception of non-human rights, or what we may rather sweepingly label environmental rights. This issue is often represented as definitional in the sense that humans have altered their perception of what are usable resources over time. The whaling issue is illustrative of these ethical discrepancies. Pro-whaling groups in Japan and elsewhere insist that whales are, like any other species of mammal or plant, resources; the most influential anti-whaling nation, the United States, asserts that whales are not to be viewed as resources but as wildlife with intrinsic worth.[31] Previous to the environmental awareness that swept the United States in the 1960s, whales were regarded as resources

---

[28] W. Catton, *Overshoot: The Ecological Basis of Revolutionary Change* (Chicago: University of Illinois, 1982), 3. See also M. Soroos, "Ecology and the Time Dimension in Human Relationships", in same and D. Orr, ed., *The Global Predicament: Ecological Perspectives on World Order* (North Carolina University, 1979), 337-343; E. Brown Weiss, *In Fairness to Future Generations: International Law, Common Patrimony and Intergenerational Equity* (Dobbs Ferry, NY: Transnational, 1989); A. D'Amato, "Do We Owe a Duty to Future Generations to Preserve the Global Environment?", *American Journal of International Law* 84 (1990); and A. Tough, "What Future Generations Need From Us", *Futures: The Journal of Forecasting, Planning and Policy*, 25:10 (1993), 1041-1050.

[29] For a discussion on related legal norms that appeal to the interests of future generations, or our "temporal neighbours", see C. Stone, "The Law as a Force", *op. cit.*, 76-77.

[30] A. Springer, "International Environmental Law After Rio: The Continuing Search for Equity", *Ethics and International Affairs*, 7 (1993), 115-130, 117. See also D.M. Johnson, "The New Equity in the Law of the Sea", *International Journal*, 31 (1976), 79-81.

[31] A. D'Amato and S. Chopra, "Whales: Their Emerging Right to Life", *American Journal of International Law*, 85:1 (1991), 21-62. We might note, however, that activities such as whale watching are usually referred to as the *non-consumptive utilization* of the mammals.

(and, in fact, they were once viewed primarily as energy resources). The debate extends to entire ecosystems. Should selected areas of rain forest in Central America, the Amazon, Southeast Asia and elsewhere be saved from deforestation in order to preserve these spaces as they are, or as future warehouses for genetic material? Do river ecosystems, which are radically altered with the introduction of damming and irrigation projects, have an inherent right to survival? Obviously, contrasting perspectives on such ethical questions will compete for our attention and devotion.

IV. SOVEREIGNTY AND THE INTERNATIONAL SYSTEM

Finally, we must mention the quandary produced by the nation-state system itself. As mentioned in our previous discussion of different perspectives on world politics (Chapter Two), different theories put different emphasis on the concept of sovereignty: some see it as the defining characteristic of states, some as a permeable institution which allows governments and transnational actors great flexibility, some as little more than another hindrance in the quest for world order or justice. In 1962, The General Assembly adopted a resolution which referred to the "inalienable right" of all states to freely "dispose of their natural wealth and resources.[32] Malaysia made particular reference to this precept during the Forestry negotiations leading up to UNCED, and this is a common North-South sticking point as southern elites claim any global environmental agenda infringes on their sovereignty.[33] Principle Two of the Rio Declaration asserts that "[s]tates have ... the sovereign right to exploit their own resources pursuant to their own environmental and developmental policies."

No matter how one views the implications of this, it is phantasmic to deny the confusing existence of sovereignty as an institution, and each of our case studies will demonstrate this. Perhaps Ervin Laszlo is correct when he writes that decision-making can grow to the international dimension if, and only if, the sacrosanct nature of national sovereignty is

[32] General Assembly Resolution 1803 (XVII), December 14, 1962.

[33] See R. White, "Environmental Management and National Security: Some Issues From Senegal", *International Journal*, XLV (1990), 106-137; J. Goldemberg, and E. R. Durham, "Amazonia and National Sovereignty", *International Environmental Affairs*, 2:1 (1990), 22-39; and J. Daudelin, "L'environnement comme Cheval de Troie? Le cas de l'Amazonie breesilienne", paper presented to the CPSA, June 1993, Carleton University, Ottawa.

surrendered. A full recognition of the historical obsolescence of that principle is the single greatest precondition for recovering the governability of contemporary societies in regard of two the most pressing issues of our times: world security, and biospheric sustainability.[34] If he is, we are in a lot of trouble. If, on the other hand, what we are seeing is the transformation of sovereignty, hope remains.

Most analysts, of course, swim in midstream. The highly respected Lynton Caldwell, who has written what is probably the most authoritative volume on international environmental policy, believes that the "jurisprudential doctrine of 'national sovereignty' is the most resistant barrier to international environmental cooperation"; but this does not lead him to conclude that the only hope for planetary survival is by way of a radical reorganization of the world's political authority structures. He insists that in order to achieve practical results the divisions and antagonisms of international society must be taken as givens.[35] Bruce Byers prescribes the "modification" of state sovereignty, and not its abandonment, "in order to recognize ecogeographical realities."[36] He does not, unfortunately, provide a concrete program for doing this, but the idea is a fascinating one, reflective of ecological considerations, and we shall return to it in our discussion of international organizations.

The debate over whether to accept environment into the cherished circle of national security concerns should not be reduced to a polemical dispute between environmental activists (many of whom know little about international relations) and military nationalists (who may know even less about the environment). The nation-state is not going to crumble and governments, influenced by ruling elites, will continue to exercise the vast majority of formal political power in the world. The danger is that the earth will crumble around the state; and that in this process anachronistic institutions will be unable to cope with this calamity. This does not deny the importance of other social agents in the process of transformation, especially in this "bifurcated and multicentric" world. James Rosenau has recently written of the "bifurcation of global structures" into the old state-centric world and the relatively ascendant multicentric world, composed of "sovereignty-free" actors including multinational corporations, ethnic minorities, subnational governments and bureaucracies, professional societies, and transnational organizations.

---

[34] E. Laszlo, "Cooperative Governance", *World Futures*, 31 (1991), 215-221, 221.

[35] L. Caldwell, *op. cit.*, 311 and 326.

[36] B. Byers, "Ecoregions, State Sovereignty and Conflict", *Bulletin of Peace Proposals*, 22:1 (1991), 65-76, 71.

This has been said before, in different ways, but the concept's resiliency is testament to the fact that the sanctified nation-state no longer dominates the political arena. It is a strained construction, unable to house all its tenants, and the question of legitimacy, which for so long has hounded international organizations, is no stranger to its door.[37] Yet it persists. Despite the increased interdependence of the modern age, the worst case scenario, a hangover from mercantilism, remains. Multiple crises could exacerbate the parochial tendencies inherent in the modern nation-state system and, facing possible ecocastrophe, nations, or ruling elites, may turn inward and attempt the impossible: self preservation only.

So if we wish to address GES, two questions we must ask are what environmental problems will lead to conflict amongst states, and what conflicts among states and sub-state actors will lead to environmental problems.[38] Methods of mitigation for either may often coincide. This does not deny the ontological attraction of globalism, which is quite evident in the environmental issue-area. But it acknowledges the realities of an international system which, for better or worse, includes states and state-actors as well as industrialists, social protest groups, economic structures, and common spaces. GES must by definition involve all these factors, and can afford to ignore none.

3. Discussion: GES and World Politics

Our inescapable, if often invisible, links with the earth unite us in a collective survival imperative. These modified and evolving ecological systems are biophysical parameters, the essential context and limit of equally complex systems of human activity. Where the two structures meet is ecology, and the effort to regulate, to mitigate the excesses of, and pursue the multiple benefits of this nexus point on a global basis is the international organization of global environmental security.

If an expanded concept of security is to be more than an empty rhetorical hull, it must promote analyses of contemporary insecurities

---

37 For a concise summary see Rosenau's article, "Normative Challenges in a Turbulent World", *Ethics and International Affairs*, 6 (1992), 1-20. A (much) lengthier exposition is found in his *Turbulence in World Politics: A Theory of Change and Continuity* (Princeton: Princeton University Press, 1990).

38 To quote Robert Cox: "A valid paradigm for the investigation of global change would need to include the historical interaction of human organization with the other elements in nature." "Multilateralism and World Order", *Review of International Studies* 18 (1992), 178.

affecting international relations and ecosystems in general. The case studies in this book will, as mentioned above, focus on three threats to ecosystem health: the spread of radioactivity (nuclear weapons and power); the loss of species (whales); and developmental environmental alteration (the Mekong River). But at this point it will be demonstrative to discuss briefly several other, broader examples. All these areas are to some degree interrelated.

Figure 3.2: A Framework for the Study of GES
THREATS to ecosystem health Threats: resource extraction and transportation; greenhouse gases and global warming; radiation; loss of biodiversity and rise of species extinction; toxic waste; soil degradation and erosion; stratospheric ozone layer depletion; marine pollutants; rapid deforestation; etc.
DECISION-MAKING PATTERNS: risks, interests, power Actors: politicians, transnational and national corporations; environmentalists; consumers of resources; international organizations; etc. Risks of threats to ecosystem: calculations based on "scientific knowledge" and interests; decision-making power and centralization and democracy; precautionary principle.
NORMATIVE CONTEXT: human and environmental rights Right of access to environmental security: secular (contemporary) and temporal (over time) equality; North-South and transnational class dimension; non-human rights of nature; future directions and competing ethics, such as free trade or conceptions of "progress".
THE INTERNATIONAL SYSTEM: sovereignty, international organization, conflict, etc. Structures: How do the above factors influence, and how are they influenced by, the international system? Is sovereignty transforming? Are international organizations gaining in influence? Is war more or less likely? Are the global structures (natural and economic and institutional) affecting the environment beyond repair?

To avoid charges of misanthropic determinism, it is worth noting that all of these developments may lead to enhanced international cooperation as well as conflict. Neo-liberal institutionalists would no doubt argue this is as likely as it is necessary. As our case studies will demonstrate, there is always this potential, and international organizations remain viable forums for such change. But they have limits which both the state-system and the transnational economic and social systems often exceed.

I. MILITARY CONFLICT

One of the lesser documented areas in the historical study of warfare is the environmental consequence of protracted military conflict. Strategic tactics, offensive and defensive, by definition involve a good deal of what we might term environmental assault and modification. This is a truism as old as war itself. As early as 2400BC, Entemenar -- the ruler of Sumer -- modified the land by constructing a canal which diverted water from the Tigris to the Euphrates watershed, which ceased Sumer's dependence on the Kingdom of Umma. As Fred Roots has pointed out, the groundwater rose with this construction, which caused rapid salinization, impoverishing Umma. Eventually, however, Sumer suffered as well, as its own over-irrigated desert soils were leached. Roots concludes that by 2200BC, "mighty Sumer was easy prey for upstart Babylon, which had less wealth and poorer technology but a clean environmental base."[39]

Arthur Westing has compiled a list of major ecologically disruptive wars which stretches back to the Persian-Scythian War of 512BC. As the Scythians retreated they hindered Persian pursuit with a scorched earth policy, a theme found frequently over the long years of military history. The Peloponnesian War (431-404BC), made famous by the account of Thucydides, saw the annually repeated destruction of Athenian grain

---

[39] E.F. Roots, "International Agreements to Prohibit or Control Modification of the Environment for Military Purposes: an Historical Overview and Comments on Current Issues", in Bruno Schiefer, ed., *Verifying Obligations Respecting Arms Control and the Environment: A Post Gulf War Assessment* (University of Saskatchewan, 1992), 13-34, 13. Roots notes that the most famous Biblical example of environmental modification for military purposes was the parting of the Red Sea by Moses: "The mechanism by which Moses accomplished this rapid environmental modification is not clear to ordinary mortals today, but presumably he did it all by triggering tectonic movements. Certainly the geological structure and accumulated crustal stress in the Red Sea graben makes this a good potential location ... if God is on your side" (ibid).

crops by the Spartans. At the end of the Third Punic War, the Roman victors polluted the farmland around Carthage with salt. Genghis Khan, leading the Mongols through Asia and eastern Europe, killed all unappropriated livestock and destroyed irrigation works located along the Tigris River in Mesopotamia. The Dutch flooded their own land to keep away French troops in the Franco-Dutch War of 1672-1678. Destruction of agricultural land, intended to starve rebellious states into the Union, was routine policy in the American Civil War. The Chinese used scorched earth tactics to put down the Tai Ping revolt (1850-1864). The Portuguese used herbicides to destroy crops during the Angolan War of Independence (1961-1975).[40]

It is clear, then, that rather than a simple "externality", or "collateral damage", environmental destruction during war is often the result of deliberate policy and strategy. The damage inflicted on Indochina's forests by American armed forces, especially the air force, provided investigative focus in the early 1970's.[41] Localized ecosystem destruction, intended to flush out and deny the supply of food to enemy forces in high-cover areas such as forests, marshes or grasslands, is best achieved by a narrow range of special-purpose weapons, especially chemical herbicides such as the infamous *Agent Orange* (2,4-D and 2,4,5-T).[42] In addition to the landscape alteration wrought by the infamous "Rome plough", more than 17,000 square kilometres of south Vietnam were damaged by herbicides, over 1500 square kilometres of ecologically sensitive mangrove forest completely destroyed.[43] This "biocide" affects land, and economic activity, far beyond the forested area itself.

Some types of environmental modification technology are much more ambitious: the American military attempted cloud-seeding

---

[40] A. Westing, *Warfare in a Fragile World: Military Impact on the Human Environment* (London: Taylor and Francis, 1980), 14-19.

[41] For early examples of such work which do not attempt to hide an anti-war bias, see J. Schell, *The Military Half: An Account of Destruction in Quang Ngai and Quang Tin* (New York: Knopf, 1968); B. Weisberg, ed., *Ecocide in Indochina: The Ecology of War* (San Francisco: Canfield, 1970); and J. Neilands, et. al., *Harvest of Death: Chemical Warfare in Vietnam and Cambodia* (New York: Free Press, 1972).

[42] J.P. Robinson, "The Effects of Weapons on Ecosystems", *United Nations Environment Programme Studies, Volume One* (Toronto: Permagon, 1979), 15. For a legal and sociological discussion of the reactions of American Viet Nam Veterans to Agent Orange, see J. Jacobs and D. McNamara, "Vietnam Veterans and the Agent Orange Controversy", *Armed Forces and Society* 13:1 (1986), 57-80.

[43] *Ibid.*, 44; also, *Report of the Special Mission on International Assistance for the Reconstruction of Viet Nam*, appointed by the Secretary-General in accordance with General Assembly Resolution 32/3 (UN Doc. 78-02689).

techniques to increase rainfall in the Indochina region in order to slow down enemy movements by inducing landslides along roadways and washing out river crossings.[44] According to one report, the U.S. actually intended to induce drought in Cuba.[45] The results of such engineering are, at best, extremely difficult to measure: though there was heavy flooding in Viet Nam in 1971, it is impossible to determine the exact cause. Critics of the Pentagon must have winced when they heard Assistant Secretary Doolin's response to charges that American forces had caused this massive flooding: "The flooding in Viet Nam, as you will recall, generated widespread civilian suffering and that was never the intention nor the result of this programme."[46] American actions in Asia were part of the stimulus which led to the signing of the *Convention on the Prohibition of Military and Any Other Hostile Use of Environmental Modification Techniques,* popularly known as ENMOD. And, though it is difficult to avoid the impression that it was put in as a last minute token gesture, Principle Twenty-Four of the Rio Declaration declares that "[w]arfare is inherently destructive of sustainable development. States shall therefore respect international law providing protection for the environment in times of armed conflict and cooperate in its further development, as necessary." We will discuss the nuclear dimension of military preparation in the next chapter.

---

[44] It became clear, "in a macabre way", writes Georg Breuer, that cloud seeding had occurred in Viet Nam, Laos and Cambodia when "the Weather Engineering Corporation [WEC] of Canada and their affiliates in the USA accused the American government of having used a special cartridge developed by their firm for seeding operations in Southeast Asia, thereby infringing the patent law, and [WEC] demanded payment of $95 million in licence fees." G. Breuer, *Weather Modification: Prospects and Problems,* translated by H. Morth (Cambridge University Press, 1979), 71-72.

[45] The idea here, according to Lawrence Juda, was to "cause clouds to drop their rain before reaching Cuba thus causing a damaging drought" in 1969-1970. L. Juda, "Negotiating a Treaty on Environmental Modification Warfare: the Convention on Environmental Warfare and Its Impact Upon Arms Control Negotiations", *International Organization* 32:4 (1978), 975-992, 976, n3. Juda is referring to a claim made by a former Department of Defense consultant, Lowell Ponte, in the *International Herald Tribune,* 29 June 1976, page 2. See also G. MacDonald, "Geophysical Warfare: How to Wreck the Environment", in N. Calder, ed., *Unless Peace Comes* (London: Allen Lane, 1968), 165-184.

[46] Quoted in Breuer, *op. cit.*, 73.

II. TRANSBORDER POLLUTION

The Chernobyl disaster provides an extreme example, though one which could very well be repeated, of the physical threat posed by transborder pollution.[47] The issue itself first gained prominence in the late 1960s, when Sweden introduced evidence that extraneous acid rain was destroying its forests. The standard North American case studies have been the ongoing Canadian-American acid rain confrontation and Great lakes pollution problems, both of which have generated inconclusive transgovernmental co-ordination.[48] The rapid fall of state socialism in Eastern Europe, and the sad state of the environment there, has brought with it an opportunity to address an even greater regional threat in Europe.[49]

The citizens of drought-stricken Africa, where environmental degradation threatens no other region in such a direct manner, should not expect the type of response Eastern Europe should solicit from the west in the coming years. Transborder pollution in much of the southern atmosphere does not so much involve emissions from industrial activity as it does water pollution in areas where two or more nation-states share water supplies. This may be expanded to include development projects -- principally, irrigation and damming projects -- which have served to exacerbate underlying tensions elsewhere. This will be discussed in Chapter Six, when we look at the Mekong River in Southeast Asia. However, if we exaggerate the term somewhat, increased pollution by China or India represents a very large transboundary pollution problem which those concerned with global warming trends insist threatens all our inter-

---

[47] A recent low-level leak of radioactive inert gases and iodine from a reactor near St. Petersburg has accentuated the fear that there are a number of Chernobyls waiting to happen. For information on this leak as well as an interesting treatment of the problem in general see D. Fairhall, "Danger Blowing in the Wind", *The Manchester Guardian Weekly* 146:13, (March 29, 1992), 1 and 12; and M. Wald, "Eastern Europe's Reactors Don't Seem so Distant Now", *The New York Times*, Oct. 13, 1991, E4.

[48] See J. Schmandt, et. al., ed., *Acid Rain and Friendly Neighbours: The Policy Dispute Between Canada and the United States*, Revised Edition (Duke University Press, 1988); D. Munton, "Dependence and Interdependence in Transboundary Environmental Relations", *International Journal*, 36:1 (1981), 139-184. The Swedish Environmental Protection Agency publishes *Enviro: Magazine of Transboundary Pollution.*

[49] See H. Schreiber, "The Threat From Environmental Destruction in Eastern Europe", *Journal of International Affairs* 44:2 (1991), 359-392, 371. On environmental problems in the area see C. ZumBrunnen, "The Environmental Challenges in Eastern Europe", in I. Rowlands and M. Greene, ed., *Global Environmental Change and International Relations* (London: Macmillan, 1992), 88-121.

ests. Indeed, the advent of ozone layer depletion and global warming, due to the widespread effect of such trends, may redefine our traditional definition of transboundary pollution in general.

Our chapter on the Mekong will touch on the issue of transborder pollution, since downstream riparian states are the most vulnerable to river pollution generated by the commercial agriculture and rapid industrialization occurring in Southeast Asia today. And whales, and all other marine mammals, remain vulnerable to transborder aquatic pollution. The hope here is a growing consensus that nation-states are responsible for the externalities (to employ the economists' phrase) of their industrial and agricultural production. The fear is that, in the absence of international law which actually holds states accountable, this consensus will fail to materialize as countries continue the rush towards industrial development.

III. RESOURCE SCARCITY

Conflict over resources may be a tired historical theme, but it is one with a bright future. This conflict may be over the right to consume resources because of their perceived scarcity; or it may be about conflict over the ethical premises of consumptionary habits, such as the rift between whaling and non-whaling nations today (see Chapter Five). As mentioned above, and as the chapter on the Mekong River makes clear in this study, there is great potential for conflict or cooperation over the most basic of resources, water. Analysts in industrial states have been more concerned with an eventual scarcity[50] of other natural resources, and the recent example of Iraq's invasion of Kuwait will heighten such apprehension. Fears of 'resource wars' involve conflict amongst powerful industrial states (usually superpowers); weaker -- perhaps allied -- states seeking the redistribution of the world's wealth; and the conflict that often results when multinational companies engage in large-scale extraction.[51]

---

[50] Scarcity is itself a very problematic concept which we have no space to delve into here; we are concerned primarily with physical scarcity as a result of unsustainable resource extraction or access denial on global terms, though discussions of scarcity within the context of a single economy or, even, mode of production, are quite interesting. On the development of the concept of scarcity as an invention of early capitalism in Europe, legitimated by the philosophic writings of David Hume, Adam Smith and others, see N. Xenos, *Scarcity and Modernity* (London: Routledge, 1989).

[51] For a succinct overview see R. Mandel, "Transnational Resource Conflict: The Politics of Whaling", *International Studies Quarterly*, 24:1 (1980), 99-127, 102-103.

The first scenario solicited much anxiety during the Cold War. A fourth fear of conflict over resources, with many origins, is that internal demands such as rising material expectations increase the aggregate demand for resource consumption and, as Choucri and North hypothesized, one rather extreme process of lateral pressure is military conquest.[52]

This is the least novel aspect of security studies. Most interstate and intrastate conflict which is based on territorial disputes involved natural resources. Westing has compiled an appendix of wars and skirmishes involving natural resources, including the two world wars, and some civil wars as well.[53] This is the stuff of modern geopolitics; but it has also been suggested that perceived land shortages lay at the root of all wars of aggrandizement in history, and that, for example, the Greeks initiated the Trojan war (1200BC) in order to assure access to tin (used to manufacture bronze).[54] Of course, it should be noted that studies of imperialism have asserted this causal link -- between pressures and demands in a core, territorial conquest of a periphery, and extraction of the latter's resources -- all along. The Europeans that set out to conquer and cultivate the world were not taking extended vacations, and Lenin's thesis regarding the origins of the First World War is well known. Regardless of whether one believes the flag follows capital or vice versa, it is clear the two have an intimate relationship. But resource war literature was not concerned with the dynamics of imperialism as much as with the possibility that the two superpowers would one day be forced to battle over strategic materials. Though most analysts agree the resource war threat was a product of anxious thinking, this has led to studies of military, political, and economic policy choices regarding the security of access to strategic resources, principally between autarkic and dependent paths.[55]

---

[52] N. Choucri and R. North, *Nations in Conflict* (San Francisco: W.H. Freeman, 1975). See also D. Orr, "Modernization and Conflict: The Second Image Implications of Scarcity", *International Studies Quarterly*, 21 (1977), 593-618.

[53] Appendix II in *Global Resources and International Conflict*, op. cit., 204-210. The list includes the Chaco War of 1932-35, the Algerian War of Independence of 1954-1962, the Congo Civil War of 1960-64, the Third Arab-Israeli War of 1967, the Nigerian Civil War of 1967-70, the El-Salvador-Honduras War of 1969 (the "Soccer War"), the Anglo-Icelandic "Clash" of 1972-73 (the "Cod War"), and the Falkland/Malvinas Conflict of 1982. We may assume the recent conflict in Kuwait would be added to this list.

[54] P. Colinvaux, *Fates of Nations: A Biological Theory of History* (New York: Simon and Schuster, 1980); D. Ager, "Ore That Launched a Thousand Ships", *New Scientist*, 106:1461 (1985), 28-29.

[55] According to Ronnie Lipschutz, military strategies to pursue internal or external autarky include the establishment of an expanded sphere of influence (the Third Reich); political strategies include isolationism (the Khmer Rouge in Kampuchea) and ideological blocs (USSR); economic strategies include import substitution (Brazil) or customs unions (EEC).

Strategic resources are of course essential to a military industrial complex. Though he refrains from accepting "an immediate causal link", David Haglund agrees with others that German mineral acquisition designs were one of the causes of World War II.[56] The great lesson of the First World War, writes Haglund, was that "modern wars of attrition would be won by those states that had the most secure resource bases, and lost by those that did not."[57] There are other reasons why governments, or transnational industrial elites, want secure resource bases. The significance of modern resources is primarily economic: the machine must be fed. We can however predict that geopolitics will resume its leading role within strategic studies as global consumption rises, global resources dwindle, and analysts argue over whether either trends actually exist.

Meanwhile, it is somewhat unsettling that environmental problems may gain legitimacy in some eyes only if they can be given a national security dimension. The current debate over resource wars and their likelihood is an example: surely, there are other reasons to pursue conservation irrespective of whether or not resource depletion may lead to actual warfare. Most literature concerned with raw materials stresses the question of the vulnerability of industrialized states to supply disruptions. This is certainly a national security question, but this misses the type of problem those concerned with environmental security from a globalist viewpoint must deal with, namely the environmental security of others. Indeed, scarcity may well be the invention of pessimists, financial opportunists, or even those with authoritarian political agendas like William Ophuls. But the relevant question with regard to mineral acquisitions by industrialized states is not supply, but, as John Young writes:

> Can the world afford the human and ecological price of satisfying its voracious appetite for minerals? Today's low mineral prices reflect only the immediate economics of extraction: purchases of equipment and fuel, wages, transportation, financing, and so on. They fail to consider the full costs of devastated landscapes,

---

Dependent policies pursue a liberal trading order, and may clash with autarkic strategies, creating war. Today, however, in an age of interdependence, many analysts doubt this possibility. *When Nations Clash: Raw Materials, Ideology, and Foreign Policy* (New York: Ballinger, 1989). See also I. Lesser, *Resources and Strategy* (London: Macmillan, 1989).

56 D. Haglund, "The New Geopolitics of Minerals: An Inquiry Into the Changing International Significance of Strategic Minerals", in same, ed., *The New Geopolitics of Strategic Minerals: Canada and International Resource Trade* (Vancouver: University of B.C., 1989), 3-34, 18.

57 *Ibid.*, 20.

dammed or polluted rivers, the squalor of mining camps, and the uprooting or decimation of indigenous peoples unlucky enough to live atop mineral deposits.[58]

IV. OVERPOPULATION

Population is a controversial ecological subject at the best of times, and perhaps more so when we treat it as a variable contributing to conflict. Hitler's use of a Lebensraum platform to justify his expansionist tendencies[59] connects population and nationalism, but more contemporary concerns revolve around the thesis, advanced chiefly by ecologists in the 1970s, that overpopulation may lead to war between neighbouring nations with migrant or refugee problems. In its simple form such a thesis is too linear as well as apolitical. No doubt, unexpected concentrations of people place increased strain on resources and social tension is likely to accompany them. But is overpopulation itself a cause? William Durham asks this question of the El Salvador-Honduras "Soccer War", often cited as an illustration of this wisdom. Demographically, El Salvador's population density is indeed alarming, and El Salvador invaded Honduras in 1969 when the latter began expelling migrants. Looking beyond generalities at the peculiar political factors which led to the war, Durham found that land scarcity in Honduras had less to do with Salvadorean immigrants than the desires of large landowners to expand commercial agriculture in the area. In this context, common in much of the agrarian world, conflict was "a response to relative rather than absolute resource scarcity."[60]

---

[58] John Young, "Mining the Earth", in L. Brown, ed., *State of the World: 1992* (New York: Norton, 1992), 100-118; 100.

[59] In 1927 Hitler told a crowded hall that sixty-two million Germans lived in an area only 450,000 kilometres square. Thus Germany was forced to either decrease the population by "chasing our best human material out" or "bring the soil into consonance with the population, even if it must be done by war. This is the natural way which Providence has prescribed." Quoted in J. Toland, *Adolf Hitler* (New York: Doubleday, 1976), 225. For an early and more general treatment see F. Hankins, "Pressure of Population as a Cause for War", *Annals of the American Academy of Political and Social Science*, 198:1 (1938), 101-108.

[60] W. Durham, *Scarcity and Survival in Central America: Ecological Origins of the Soccer War* (Stanford University Press, 1979), 164. Durham lays the blame on special interest groups: the National Federation of Agriculturalists and Cattle Ranchers of Honduras "succeeded in translating an internal problem of resource competition into an external one." *Ibid.*, 165. See also T. Homer-Dixon, *op. cit.*, 82.

Fears of the southern hordes are already driving the post Cold War agenda of Northern insecurity. The general argument is that, given present demographic trends, the "Third World", already a clear majority, will become an overwhelming one, threatening the North's sense of security and, even, values. Eberstadt warns that in this world of over 10 billion, "[even without the rise of new blocs or alignments, one can envision a fractious, contentious and inhumane international order: liberal precepts could have steadily less impact on international action and belief in human rights could prove a progressively weaker constraint on the exercise of force."[61] These long-term scenarios, partly the product of paranoia, partly the product of slavish belief in the inevitability of demographic trends, and partly the result of a genuine concern with overpopulation as an ecological problem, lead to a seige mentality might even interpret increased Third World populations as a means to aggression on the part of the south, and not a cause.[62] In other words, if we start with the premise that nations in the southern hemisphere are characteristically aggressive and will take what the North now cherishes the first chance they get, increased populations there will be an obvious threat to global security.

We should be careful before adopting such a viewpoint. Both population density and migrant or refugee expulsion leads to problems (witness relations between Ghana and Nigeria in the early 1980s); but, as Nazli Choucri points out, there are a host of intermediary factors -- such as the roles of technology, the state, and ideology -- which "relate and link conflictual outcomes to demographic factors."[63] There is no simple, straight paths in the dense forest of social science.

---

[61] Eberstadt, *op. cit.*, 129. See also B. Wattenberg, *The Birth Dearth* (New York: Pharos, 1989); G. Foster, "Global Demographic Trends to the year 2010: Implications for U.S. Security", *Washington Quarterly* 12 (1989), 5-24; and C. Kegley and E. Wittkopf, *World Politics: Trend and Transformation, Fourth ed.* (New York: St. Martin, 1993), 317-318.

[62] We should note that Eberstadt is aware of the problems associated with using demographic trends as the basis of analysis.

[63] N. Choucri, "Perspectives on Population and Conflict", in same, ed., *Multidisciplinary Perspectives on Population and Conflict* (Syracuse: Syracuse University Press, 1984), 1-25, 19. For a general treatment of population and international affairs, and an extensive bibliography, see M. Leroy's *Population and World Politics: The Interrelationships Between Demographic Factors and International Relations* (Leiden: Netherlands Interuniversity Demographic Institute, 1978).

V. Climate Change

Long-term threats to national security posed by anthropogenic global environmental alterations are multiple and all contribute to global insecurity. A great deal of scientific and political uncertainty characterizes this area. Though he is ambiguous, David Wirth believes that unchecked global warming, with its potentially negative consequences for American agricultural capacity, could exacerbate the risk of war as it produces major shifts in the international balance of power.[64] Even if one accepts balance of power models which predict hegemonic war, this is heady stuff, rooted partly in the dubious assumption that present power configurations are solely the result of favourable climate.[65] However, as Wirth points out, the security implications of continued greenhouse warming to countries with a large proportion of land barely above sea-level, such as the archipelagic Maldives, Guyana, the Netherlands, and many Pacific islands, is obvious, and the ensuing environmental refugee problem will be particularly stressful where heavily populated areas such as Bangladesh are affected (there already exists an acute situation in Assam, India, though this is related to deforestation). Furthermore, current geographical fixtures which enhance physical sovereignty, such as ice in the Arctic, may be eroded; for example, the Canadian Government may have a difficult time asserting its claim to sovereignty over 'internal' arctic waters if extra-national access to them resulted from greenhouse warming.[66]

---

[64] D. Wirth, "Catastrophic Climate Change", in M. Klare and D. Thomas, ed., *World Security: Trends and Challenges at Century's End* (New York: St. Martin's, 1991), 381-400, 388. One analyst, Francis Bretherton of the U.S. National Center for Atmospheric research, suggests that as the Great Plains turn into a dust bowl, America's people and wealth will move inexorably northwards, and Canada could become one of the worlds most powerful nations. This is a fascinating scenario, though it is more likely that Canada's increasing attractiveness to the United States would have serious implications for the former's sovereignty. Bretherton is quoted in F. Osler Hampson, "Peace, Security, and New Forms of International Governance", in C. Mungall and D. McLaren, ed., *Planet Under Stress: The Challenge of Global Change* (Toronto: Oxford, 1990), 301-317, 309. For more on the possible effects of climate change, see I. Mintzer, ed., *Confronting Climate Change: Risks, Implications and Responses* (Cambridge: Cambridge University Press, 1992); and M.E. Fernau, et. al., "Review and Impacts of Climate Change Uncertainties", *Futures: The Journal of Forecasting, Planning and Policy*, 25:8 (1993), 850-863.

[65] This is the line of reasoning adopted by, among others, S.F. Markham, in *Climate and Energy of Nations* (London: Oxford, 1944), and C.A. Mills in *World Power and Shifting Climate* (Boston: Christopher, 1963).

[66] The *Globe and Mail* reports that "if warming of the world's climate melts Arctic sea ice, Canada may have to take steps to assert its sovereignty over the Northwest Passage"; Ian

But the people most immediately affected by climate change will be those already living in conditions where daily survival is daunting, and the rapid increase in "environmental refugees" has been widely predicted.[67] In the short term, those nations, or classes within nations, with adequate infrastructure and mobile wealth will, perhaps with great difficulty, be able to adapt, while those unable to do so may, in an increasingly hostile and parochial world, be left to their own misfortune. If drought and desertification cannot be addressed now, will the future bring about a moral or technological improvement? And this will be exacerbated by the resource depletion induced by what can only be described as another important threat to global environmental security, Third World debt.[68]

Global warming presents a classic example of uncertainty affecting decision-making, since it is unpredictable and thus challenges advocates of the precautionary principle to explain how to operationalize the ethic. Since major cut-backs in carbon emissions would have serious economic implications at this point in time, especially for some key industries around the globe, this will probably not become anything more than an academic problem in the near future: the 1992 Framework Convention on Climate Change signed at Rio was predictably disappointing to those -- and the number seems to be increasing -- who consider climate change the number one long-term threat to humanity.[69]

Jackson writes for the Institute for Research on Public Policy in Ottawa that "[a]t present, ice provides the effective barrier on the Earth's surface that protects Canadian sovereignty in the North." J. Rusk, "Global Warming Could Melt Sovereignty", *Globe and Mail*, Tuesday, July 21, 1992. The Canadian military is taking this possibility seriously and has commissioned work on potential scenarios. See for example C. M. Tucker, *Environmentalism and Environmental Change as it Affects Canadian Defence Policy and Operations*, "Part Three: Arctic Issues"; Operational Research and Analysis Establishment Report No. R106 (Ottawa, 1990), Microlog 91-01753.

[67] See G. Woodwell, "The Effects of Global Warning", in Jeremy Leggett, ed., *Global Warming: The Greenpeace Report* (Oxford: Oxford, 1990); C. Tickell, "Environmental Refugees: The Human Impact of Global Climate Change", in T. Minger, ed., *Greenhouse Glasnost*, op. cit., 181-194; H.J. Karpe, et. al., eds., *Climate and Development: Climatic Change and Variability and the Resulting Social, Economic, and Technological Implications* (Berlin: Springer-Verlag, 1990); and Chapter Seven of this study.

[68] See M. Miller, *Debt and the Environment: Converging Crises* (New York: United Nations, 1991); F. Cheru, "Structural Adjustment, Primary Resource Trade and Sustainable Development in Sub-Saharan Africa", *World Development*, 20 (1992), 497-512; N. Hopkinson, *Desertification, Debt and Structural Adjustment in Sub-Saharan Africa* (West Sussex: Wilton Park, 1989); A. Isla, "The Debt Crisis in Latin America: An Example of Unsustainable Development", *Women and Environments*, 13 (1993), 34-36; and J. Cartwright, "Conserving Nature, Decreasing Debt", *Third World Quarterly*, 11:2 (1989), 114-127.

[69] Author's observations based on communication with international affairs analysts in Canada and elsewhere.

4. Conclusion

It gives rise to false expectations to claim international organizations, or even established patterns of cooperation between nation-states (regimes), are capable of solving what are in fact deeply embedded problems reflective of modern global society. They do, however, play a major role in defining and articulating them at the political level, especially within the United Nations system, the efficacy of which has been oscillating since its inception. This is so even with regard to traditional definitions of security: unless we rely entirely on the supposed stability provided by a balance of power in which there are no significant intervening variables between states, we must be concerned with international organization. Yet surely this is but part of the role contemporary global organization has to play. International organizations in particular have been the forum for international discourse about an increasingly wide array of problems, and environmental degradation continues its ascent as an item of great concern. And the modern saliency of non-governmental, as opposed to intergovernmental, organizations should not be forgotten either. The threats to the environmental security of contemporary earth, it is well known and accepted, do not recognize our neatly (and not so neatly) delineated world comprised of political borders and territorial sovereignty. It is imperative that the conflicts resulting from this delineation are in some manner controlled by international organization so that dealing with those threats becomes a possibility. Hannah Arendt, who once claimed the world had to choose between socialism and barbarism, may be right in the long run; but we do not have the ecological time to make this discovery. For now, the choice is less fundamental but equally significant, and it is between international cooperation, hopefully designed along the lines of some globalist conception of security, and the barbarism which could result when humanitarian responses to crises become unthinkable. This was in essence the argument advanced by William Ophuls, who was misinterpreted as a Hobbesian authoritarianist but argued that his essential political message "... is that we must learn ecological self-restraint before it is forced on us by a potentially monolithic and totalitarian regime or by the brute forces of nature. We are currently sliding by default in the direction of one (or both) of these two outcomes. Only the restoration of some measure of civic virtue (to use the traditional term) can forestall this fate."[70]

It is clear that the GES landscape is an intertwining and complex one. The ultimate task facing international organization may well be to translate normative discourse into concrete political action, and, given the co-existent contexts of global cooperation and conflict, we should be very sceptical before assuming this even possible, much less likely. Despite an obviously cautious theme, however, this study is not misanthropic in conception. Possibilities, most often born of crisis, exist. Redefining security on an ecosystem basis, though it raises a plethora of interesting and frustrating political problems, could be a limited answer.

This study will demonstrate this duality, scepticism and optimism, with reference to political activity affecting three special earthly things: atoms, whales, and rivers.

[70] Quoted in W. Ophuls and A.S. Boyan, *Ecology and the Politics of Scarcity Revisited: The Unraveling of the American Dream* (New York: Freeman, 1992), 312.

CHAPTER FOUR

ATOMS: NUCLEAR TRADE, WEAPONS, AND DISCOURSE

"We are here to make a choice between the quick and the dead."
Bernard Baruch, 1946[1]

"...Foucault suggests in *Discipline and Punish* that the power of organized societies is also generative. It causes people to act in certain ways, not only by the limited means of coercion, but by creating a field in which various actions are privileged ... "
Christopher Manes, 1990[2]

1. INTRODUCTION

The primary assumption upon which this chapter is based is that the spread of both nuclear weapons and nuclear power installations constitute principal threats to global environmental security. It is of course widely accepted that even local nuclear wars, if such limited conflicts are even possible, would be nothing short of ecological disasters. Indeed, scientific inquiry into the possibility of nuclear winter helped put environmental security onto the international agenda. Even in an era of disarmament, concern about potential nuclear weapons accidents remains

[1] From Baruch's speech to the First Meeting of the Commission to Deal With the Problems Raised by the Discovery of Atomic Energy and Other Related Matter, Hunter College, The Bronx, New York, 14 June 1946 (text in UNAEC, Official Records, No. 1, 1946), page 4.

[2] C. Manes, *Green Rage: Radical Environmentalism and the Unmaking of Civilization* (Toronto: Little, Brown and Company, 1990), 230.

strong. Revelations continue to surface about the environmental costs of nuclear weapons production, testing, and disarmament. And the negligent management of nuclear power systems, particularly in the former Soviet Union, has given atomic energy an equally ominous stigma. The hazards of uranium mining, the threat of meltdown, and the insoluble problem of indefinite nuclear waste disposal make large-scale atomic energy incompatible with a progressive vision of sustainable development.[3] As we near the landmark of 1995, the year in which the Non-Proliferation Treaty (NPT) will be renewed indefinitely or for a specific period of time, we should take stock of what a newfound concern for environmental security at the global level means for the nuclear issue, and at what normative factors underpin both understanding and behavior in this area.

The nuclear marketplace is a unique and highly controversial arena in the world political economy and system. Is any form of international trade viewed with comparable monition? The trade is unusually regulated because of a genuine, and reasonable, fear of the horizontal proliferation of nuclear weaponry. The persistence of the nuclear industry -- indeed, its own horizontal growth despite a falling world market -- may be attributed in part to a great diplomatic effort at the multilateral level which has assumed a double agenda: preventing proliferation is the overt goal, while the legal and moral legitimation of nuclear trade has been the less dramatic (though hardly silent) counterpart. This chapter will explore the "civilian-uses" issue, notable for its presence throughout the nonproliferation regime and certain to continue its role in the upcoming negotiations towards the important 1995 NPT renewal session. Such an exploration draws needed attention to the political economy and normative dimensions of the problem of nuclear proliferation and, from an environmentalist perspective, nuclear power risks and sustainable development.

It has become standard practice to refer to the various actors and institutions involved in the international nuclear issue as part of a

---

[3] For literature related to all these factors, see C. Sagan, "Nuclear Winter and Climatic Catastrophe: Some Policy Implications", *Foreign Affairs* 62 (1983/4), 257-292; S. Gregory, *The Hidden Cost of Deterrence: Nuclear Weapons Accidents* (London: Brassey's, 1990); S. Sagan, *The Limits of Safety: Organizations, Accidents, and Nuclear Weapons* (Princeton: Princeton University Press, 1993); K. Steele, "Cleaning Up After the Cold War", in A. Hammond, ed., *The 1993 Information Please Environmental Almanac, Compiled by the World Resources Institute* (Boston; Houghton-Mifflin, 1993), 129-146; N. Lenssen, "Facing Up to Nuclear Waste", *Worldwatch* March/April 1992, 10-17; T. Halverson, "Ticking Time Bombs; East Bloc Reactors", *The Bulletin of the Atomic Scientists*, July/August 1993.

"regime." As we have discussed at some length the liberal institutionalist (or pluralist) perspective on regimes, it is unnecessary to delve into theories of regimes here. The nuclear non-proliferation regime is usually seen as comprised of a number of international agreements and mechanisms, including the NPT, the IAEA safeguards system, nuclear export guidelines and associated national controls, nuclear testing limitations, regional agreements on nuclear-weapon-free-zones, and nuclear guarantees from existing nuclear weapons states.[4] We might add the associated 'missile technology control regime' and others. Importantly, however, and as discussed earlier in this study, regimes are both prohibitionary and promotional. They act as formal constraints on the behavior of decision-makers, in private and public industry. But they also act to legitimize certain forms of behavior, and in world trade to make palatable what may in essence be quite troubling and potentially very controversial patterns of interaction. This is often obscured by the tendency to view instances of international cooperation, in the context of an anarchical order which always presents the possibility of war, as inherently progressive. A more critical perspective informs this study, as well as the others contained in this book. The nonproliferation regime also consists of normative assumptions, which we will explore below, that serve to legitimize nuclear trade in general. Environmental concerns may serve to change this, altering patterns of risk assessment.

It is conventional to view nuclear proliferation from the demand side, either in terms of the Third World's supposed "pressing nuclear energy needs"[5] or, ominously, nuclear weapons aspirations of the *threshold states* who seek to develop arsenals by way of formal or clandestine trade routes. Recent developments in Iraq and North Korea encourage this orientation. However, there is growing concern with the supply side of the contemporary market. This reverse approach includes as threats to the nonproliferation regime new and old nuclear suppliers, defined herein

---

[4] J. Simpson and D. Howlett, "Nuclear Non-proliferation: The Way Forward", *Survival*, 33:6 (1991), 483-499. See also T. McMorris-Tate, "Regime Building in the Non-Proliferation System", *Journal of Peace Research*, 27:4 (1990), 399-414; J. Keeley, "Towards a Foucauldian Analysis of International Regimes", *International Organisation*, 44:1, 1990; and G. Bertscht and R. Cupitt, "Nonproliferation in the 1990s: Enhancing International Cooperation on Export Controls", *The Washington Quarterly*, 16:4 (1993), 53-70.

[5] We may note the assumptive language nuclear industry representatives like Bujon de l'Estang utilize. While no one doubts that there are pressing energy needs in the Third World, this hardly means they are, necessarily, nuclear. F. Bujon de l'Estang, "The Delicate Balance: Governmental and Industrial Cooperation in Enforcing Nonproliferation", in R. Jones, et. al., ed., *The Nuclear Suppliers and Nonproliferation: International Policy Choices* (Lexington: Lexington Books, 1985), 131-142, 133.

as states which host firms (public and private) that trade abroad in nuclear and nuclear-related technology, equipment, materials and services. This inclusion results mainly from anxiety that, in what has with the complications of nuclear power generally become a *buyer's market*, states will lower safeguard standards in export policy in order to secure valuable trade. The disarray of the nuclear industry in the former Soviet Union increases fears that horizontal proliferation will be induced by supply-side actors, including former nuclear scientists. These market dynamics occur within the context of the often confrontational North-South dichotomy of the global system which has evolved at the diplomatic level of nonproliferation regime maintenance. Accusations from southern states of treaty negligence regarding promised nuclear arms reduction and technology transfers have been permanent vexations. Reference is made to Article 6 of the Non-Proliferation Treaty (NPT), wherein weapons states pledged in good faith to reverse the nuclear arms race; and Article 4, wherein non-weapons states are promised access to peaceful nuclear technology.

The discourse of atomic diplomacy is parochial in scope and based on several dominant assumptions. The one pertaining most especially to the civil nuclear energy question and its North-South dimension is the prevalent ideology of development itself, only slightly tempered by the advent of environment and sustainable development as a major policy concern. Shared by the policy and social elite of the South and North, and the international institutions which they influence, is the belief that underdeveloped nations "will find solutions to their problems only by following models of ... energy-and-resource intensive industrial development prescribed by the North."[6] The nuclear industry has managed to sell products (and must project future markets doing so) to nations whose prescriptive development policy choices place great hope in an industrial future, and the examples of what are now commonly referred to as NICs (Newly Industrializing Countries), such as South Korea and Taiwan, lend credence to these expectations. But the enormous costs of nuclear power, that "most capital-intensive of all technologies"[7], and public opposition to its implementation in both the North and South have combined to miti-

---

[6] L. MacKay, "Lines of Longitude: People's Detente, North and South", in M. Kalder and L. MacKay, eds., *Something in the Wind: Politics After Chernobyl* (London: Pluto, 1988), 131-160, 136.

[7] W. Walker, "Nuclear Non-Proliferation and the Present Status of the International Nuclear Industry", in J. Simpson and A. McGraw, eds., *The International Nuclear Non-Proliferation System: Challenges and Choices* (New York: St. Martin's, 1984), 91-107, 96.

gate such sales severely. An oversupply of nuclear commodities and uranium has become the norm. The great apprehension relating to both nuclear weapons proliferation and energy systems' operational safety results from the seller's incentive, in this buyer's market, to devalue safeguards before products.[8]

This chapter will present a brief review of the circumstances which have led to this predicament. We begin by describing the great expectations which aided private sector demands for the commercialization of nuclear power in the United States. This leads to a discussion of the nexus between state and industry in the nuclear field. The eventual decline of American dominance in that field precipitated the growth of what is now a global export industry. We will examine the transnational phenomena associated with this issue from the North-South perspective: the actors and, as importantly, normative factors which transcend the bounds of state diplomacy. This normative inclusion is based on the earlier argument that modern studies of environmental security issues should take normative factors into account, even if this disturbs traditional notions of 'objective social science'). Finally, we will discuss the nonproliferation regime's future -- assuming for now there is one to discuss. The diffusion of supply sources makes the management of a credible regime much more difficult, but we are bound to continue the search for universality, not despite but because of the Iraqs and North Koreas. The quest for environmental security demands it; but it also demands that we look beyond the weapons question, at the civil power question, and at the issue of sustainable development. Perhaps this is an opportune time to begin a broadened diplomatic discourse in this vital area of world politics.

---

[8] It may be noted, however, that price has in fact been affected, especially with regard to uranium. In a downward slide indicative of the contraction of the reactor industry, uranium, once priced at over $40 (U.S.) per pound, fell to less than $20 by the late 1970's (Patterson, 1984: 131) and was valued as low as $9.50 in 1991 (*The Globe and Mail*, March 21, 1991, pp. B1 and B2). By 1993 it had fallen further, to eight dollars a pound. B. Semenov and N. Oi, "Nuclear Fuel Cycles: Adjusting to New Realities", *IAEA Bulletin* 35:3 (1993), 4. As for total energy demand, the International Nuclear Fuel Cycle Evaluation (INFCE) predicted in 1980 that market economies would produce between 850 and 1200 gigawatts-electric (Gwe) with nuclear energy by the year 2000; the most recent IAEA estimate is between 372 and 382 Gwe (current global production, from 424 nuclear plants, is around 330). *Ibid.*

2. The Nuclear Predicament

Despite the attention given to the potential environmental consequences of the Iraq-U.S. (UN) war, there is a current preoccupation with the environmental problems associated with Superpower activity and, especially, nuclear-weapons production. The Cold War will not be over until the hazardous legacy of nuclear weapons production and testing in both superpower states, but particularly in the former Soviet Union -- where the not-so secret atomic cities[9] present the greatest threat to environmental security -- is disposed of. The cost of this endeavour will be enormous. A recent Russian Government disclosure, for example, admits that the giant Mayak plutonium plant, established near Chelyabinsk in 1948, has been the scene of two large accidents and, worse, the plant "dumped radioactive waste into a nearby river during the first seven years of its operation", affecting more than 450,000 people with radiation.[10] After 1951, wastes from the Chelyabinsk plant mentioned above were dumped directly in Lake Karachay, contaminating local groundwater and, according to one report, "the radiation is ... so great today that as little as one hour's exposure at the shoreline could prove fatal."[11] Parts of the lake had to be covered with concrete due to this excessive, and excessively irresponsible, nuclear waste dumping. Indeed, in the saddest of commercial ironies, cleaning up the Soviet Union has been a source of potential employment for American firms. It was announced in London on 1/31/92 and reported in the *Financial Times* that "the International Disarmament Corporation, a company that will specialize in the dismantling of Soviet nuclear weapons systems, has been formed by Lockheed, Olin, and the McDermott International subsidiary Babcock and Wilcox ... the group will offer its service to the four commonwealth states that lack the [appropriate] expertise..."

On the other side of the Atlantic, there is little doubt that the biggest polluter in the United States is the Pentagon. The military's 'toxic empire' -- more than 20,000 sites contaminated with millions of tons of toxins and hazardous chemicals -- is the subject of a new book by Seth

---

[9] As recently as April, 1993, an explosion at the formerly secret Tamsk-7 nuclear weapons production plant in Siberia occurred. "*Radioactive Cloud Scuds Over Siberia After Blast*", *The Toronto Globe and Mail*, April 8, 1993.

[10] "Nuclear Plant Dosed 450,000 in Accidents", *The Toronto Globe and Mail*, Jan. 28, 1993.

[11] P. Gizewski (Canadian Centre for Global Security), "Military Activity and Environmental Security: The Case of the Arctic", paper presented to "Environmental Security After Communism" Conference, Carleton University, Ottawa, Feb. 26-27, 1993.

Shulman.[12] Only 404 of the 20,000 sites on U.S. Defense Department land have been cleaned up. Rivalling the output of the top five U.S. chemical companies combined, the Pentagon produces well over a ton of toxic waste every minute. Though considerable expense is being taken to clean them up, former weapons plants like those at Hanford, Washington, Savannah River, South Carolina, and Rocky Flats, Colorado, have with ironic flair been termed "national sacrifice zones".[13] This is a direct legacy of the Manhattan project, which culminated in the testing of the Trinity bomb in July of 1945.

Human-nature relations have been routinely altered with technological innovations, many of them conducive to expanding the parameters of organized violence.[14] But the invention and demonstration of the atom bomb was the most significant technical development in modern history. For the first time, it became credible to conceive of an extant threat to global civilization, of omnicide itself, and subsequent calls rang out (to paraphrase Albert Einstein) for changing our mode of thinking since everything else had changed. By considering the damage-potential of nuclear weapons, scientists inevitably cast light on the possibly harmful effects of nuclear war on ecosystems. Perhaps the most widely read account of such armageddon-like forecasting was Carl Sagan's discussion in *Foreign Affairs* of the possible climatic catastrophe -- nuclear winter -- which would follow even small-scale nuclear war, though an earlier book by Jonathan Schell had a popular reaction, and public awareness of the ill effects of nuclear testing was partly responsible for the eventual accep-

---

[12] S. Shulman, *The Threat at Home: Confronting the Toxic Legacy of the U.S. Military* (Boston: Beacon, 1992). On the effects of nuclear production by the U.S. military, see U.S. Office of Technology Report, "Complex Cleanup: The Environmental Legacy of Nuclear Weapons Production" (Washington: USGPO, 1991), and Karen Dorne Steele's concise "Cleaning Up After the Cold War", in A. Hammond, ed., *The 1993 Information Please Environmental Almanac, Compiled by the World Resources Institute* (Boston: Houghton-Mifflin, 1993), 129-146.

[13] R. Alvrez and A. Makhijani, "Hidden Legacy of the Arms Race: Radioactive Waste", *Technology Review*, 91:6 (1988); 42-51. At the Savannah River plant, experts had predicted that plutonium waste would not reach the water table for 1 million years: it took only 20. *Ibid.*, 47. See also J. Ahearne, "Fixing the Nation's Nuclear Weapons Plants", *Technology Review*, 92:5 (1989), 24-29; and S. Hilgartner, et. al., "Hanford: The Interim Reality", in their *Nukespeak: The Selling of Nuclear Technology in America* (New York: Penguin, 198), 150-159. For an interesting account of America's overseas military environmental problems, see J. Walker, "A Green Peace: The Closure of Overseas Military Installations and the Environmental Liability of the U.S. Government", *Georgetown International Environmental Law Review*, 3:2 (1990), 437-456.

[14] For what has become a classic study see R. Preston and S. Wise, *Men In Arms: A History of Warfare and Its Interrelationships with Western Society,* Fourth Edition, (New York: Holt, Rinehart and Winston, 1979).

tance of the Partial Test Ban Treaty in 1963.[15] The nuclear winter debate which followed Sagan and others' work assumed familiar ideological lines, as Theodore Rueter and Thomas Kalil demonstrate in their review essay. As they write, "for hawks, nuclear winter was an additional argument for counterforce and limited nuclear options; for doves, it was an additional argument for disarmament and arms control."[16]

As for military applications, the fear of accidental radiation discharge has increased as well. Nuclear accidents in the United States military are classified, from least serious to worst, as "Dull Swords, Bent Spears, Broken Arrows, or Nucflash." Nucflash involve accidents that could lead to nuclear war, and they have so far been avoided. But the Department of Defense has acknowledged "at least 27 Broken Arrows between 1950 and 1981. All involved the Air Force. On at least two occasions when planes carrying nuclear bombs crashed, the bombs were destroyed (though they did not explode) and plutonium was scattered over a wide area."[17] Other analysts, looking at American, British, and Soviet records between 1950 and 1988, record over 190 accidents which "caused, or might have caused, detonation of or damage to a nuclear weapon." The authors conclude that there is an "inevitable contradiction between weapons safety and weapons readiness."[18] And the environmental effects of atmospheric nuclear testing were at least partially responsible for the eventual adoption of the Partial Test Ban Treaty of 1963. Regardless of ideological orientation, if there has been a change in consciousness regarding the atom, it would seem to be towards a less

---

[15] C. Sagan, "Nuclear War and Climatic Catastrophe: Some Policy Implications", *Foreign Affairs* 62 (Winter 1982/84), 257-292; J. Schell, *The Fate of the Earth* (New York: Knopf, 1982). For related, if more technical, material see R.P. Turco, et. al., "Nuclear Winter: Global Consequences of Multiple Nuclear Explosions", *Science* 23, Dec. 1983, 1283-92; P. Ehrlich, et. al., "Long-term Biological Consequences of Nuclear War", *ibid.*, 1293-96; and J. Peterson, ed., *The Aftermath: The Human and Ecological Consequences of Nuclear War* (New York: Pantheon, 1983).

[16] T. Rueter and T. Kalil, "Nuclear Strategy and Nuclear Winter', *World Politics* 43 (July 1991), 587-607, 590. See also B. Haig, "Why We Must Accept Nuclear Winter Theory", *Bulletin of Peace Proposals* 20:1 (1989), 81-88.

[17] C. Caufield, *Multiple Exposures: Chronicles of the Radiation Age* (London: Secker and Warbug, 1989), 243. See also D. Kaplan, *Nuclear California* (San Francisco: Greenpeace, 1982), 54. Caufield adds: "The Navy has also acknowledged that it has several hundred pages of documents under the heading 'Summary of Navy Nuclear Weapons Accidents and Incidents', covering the period from 1973 to 1978. The Army claims never to have had a serious accident with nuclear weapons."

[18] S. Gregory and A. Edwards, "The Hidden Cost of Deterrence: Nuclear Weapons Accidents, 1950-88", *Bulletin of Peace Proposals* 20:1 (1989), 3-26, 4.

generous attitude regarding its potential for liberating humanity from the grip of either the security dilemma or nature, though cold fusion remains an enticing -- and quite controversial -- prospect in the latter case, and proponents of deterrence remain committed to the former.

Despite the long-term threat to environmental security represented by these problems, consensus is emerging that the greatest immediate danger in both eastern Europe and the newly formed Commonwealth of Independent States is the decrepit condition of the outdated nuclear reactors located there. A recent low-level leak of radioactive inert gases and iodine from a reactor near St. Petersburg has accentuated the fear that there are a number of Chernobyls waiting to happen.[19] Throughout the Cold War, eastern Europe had served as a captive market for the Soviet nuclear industry, and Soviet-designed reactors (located in Czechoslovakia, Hungary, Bulgaria, and eastern Germany) have not always included sufficient emergency core cooling systems.[20]

Since the nuclear disaster at Chernobyl in April of 1986, there has been increased concern over the safety of Soviet-inspired nuclear technology and operations. The international community has for a variety of reasons been slow to respond, despite two mid-1980s agreements regarding the early notification of nuclear accidents and the provision of assistance in the case of nuclear accident or radiological emergency. The establishment of the WANO (World Association of Nuclear Operators) reflected part of what Martin List terms "[t]he concern that the inadequate safety of Eastern nuclear power plants might discredit Western nuclear energy programmes [which] was one of the main motivational forces for the post-Chernobyl activities of nuclear power plant operators."[21]

The European Community itself has established two general aid-for-the-east programmes: PHARE (1990) and TACIS (technical assistance to CIS states, 1991). There is also a "nuclear safety account" -- 60 million ECU -- set up at the European Bank for Reconstruction and Development (EBRD), to which Canada -- interested in the future market potential --

---

[19] For information on this leak as well as an interesting treatment of the problem in general see D. Fairhall, "Danger Blowing in the Wind", *The Manchester Guardian Weekly*, March 29, 1992, 1 and 12.

[20] M. Wald, "Eastern Europe's Reactors Don't Seem so Distant Now', *New York Times* Oct. 13, 1991, E4.

[21] M. List, "The West and Eastern Nuclear Power Plant Safety", *Aussen Politik*, 44:4 (1993), 344-353.

has contributed.[22] The spread of nuclear power to the southern hemisphere, which has so far been a very limited and troubled affair, could mean much more of the same technical negligence and reactive commitment.

3. Industrial Expansion ... and Contraction

In order to understand how nuclear thinking reached this point, we may begin by briefly recalling the ebullience which greeted the initial public discovery of atomic energy in the United States. The conquest of the atom, which ended the Second World War, represented an unprecedented victory of "big science" over nature, a technocratic milestone coincident with rapidly growing energy demand projections. This euphoric atmosphere of the 1950's was nurtured by publicized visions of nuclear powered district heating, motor vehicles, and even vacuum cleaners. These were of course largely commercial images, conjured by the American electro-technical industry, whose corporate structure predated the intense atomic research of the Manhattan Project and which lobbied for the commercialization of nuclear power.[23] Scientific consensus validated this quest for the widespread enjoyment of the atom: Novick reports that, as early as the 1940s, "highly respected individuals predicted that the atom would make electricity too cheap to meter"; Schiff writes of the "glorious visions [of] nuclear-powered agro-industrial complexes"; and Patterson of the "plutonium dreamers" enticed by the prospect that "modern alchemy could transform the otherwise useless rock of uranium ore into an effectively limitless supply of high-quality energy."[24] There

---

[22] I am grateful for a relevant discussion in January of 1994 to Roger Lucy, Programme Manager for the Canadian Nuclear Safety Initiative. Many environmentalists are critical of what they view as too conciliatory an approach to the existing problem, arguing that this should be viewed within the context of the global nuclear power marketplace, which is currently a "buyer's market": western companies see a potential market in the new eastern Europe and, if they are too harsh in condemning the present reactors, "people might not want them to build new ones." A. Froggatt, of Greenpeace, quoted in ibid.

[23] W. Walker and M. Lonnroth, *Nuclear Power Struggles: Industrial Competition and Proliferation Control* (London: Allen and Unwin, 1983), 19, 20.

[24] Respectively: S. Novick, "The Menace of the Peaceful Atom", in F. Knelman, ed., *1984 And All That: Modern Science, Social Change, and Human Values* (Belmont, Calif.: Wadsworth, 1972), 217-227, 217; B. Schiff, *International Nuclear Technology Transfer: Dilemmas of Dissemination and Control* (London: Croom Helm, 1983), 165; and W. Patterson, *The Plutonium Business and the Spread of the Bomb* (London: Wildwood House, 1984), xiii.

was also an early faith in the constructive potential of the peaceful nuclear explosion (PNE) for excavationary purposes, illustrated by Glenn Seaborg's insistence that he and other atomic officials "truly believed that the PNE program could lead to enormous and otherwise unobtainable benefits for mankind."[25] These promises were made in an ideological environment conducive to nuclear commercialization: for example, one of the government officials responsible for the latter, which began in the mid-1950's, was Lewis Strauss, a "die-hard conservative with an ideological devotion to pure private enterprise."[26]

However, nuclear energy would never completely fall into that economic realm. Mutual need was a factor. The American government would need private sector participation in order to maintain the Western nuclear dominance to which American security elites quickly became accustomed, and the private sector would need state cooperation to guarantee profitable expansion. Indeed, it is often difficult (and especially so on the international level) to distinguish between public and private sector behavior in the nuclear field as a whole since the two became so inextricably linked, a process that begun in earnest after the transfer of most civil nuclear technologies to private industry with the "Atoms for Peace" program initiated in 1953. That the nuclear industry would be heavily regulated was assumed from its beginning, and this may have been regretted by its expansionists. Yet there is little doubt that the industry, which originally demanded more governmental assistance than it received, gained in two senses from state intervention: in concrete terms, including research aid and outright subsidization; and, more subtly, in terms of the legitimation of the industry itself, which would soon become an international task.[27]

This policy isthmus between capital and state is significant in the context of export policy. Of course, it is also the subject of a polemical dispute and a question of historical interpretation. Did the "Atoms for

---

[25] G. Seaborg and B. Loeb, *Stemming the Tide: Arms Control in the Johnson Years* (Lexington, Mass.: Lexington Books, 1987), 347.

[26] M. Hertsgaard, *Nuclear Inc.: The Men and the Money Behind Nuclear Energy* (New York: Pantheon, 1983), 31.

[27] Legitimation involves legal sanction as well as less easily discernible normative support. J. Camilleri writes of an institutionalized *nucleocracy*: "In the nuclear context, the state was obliged to allocate vast resources in order to persuade the public that nuclear technology was the answer to the 'energy crises', to the fear of 'not having enough', and that only technical expertise ... offered the basis for such a solution. [This was] the ideological manipulation of perceived scarcity..." *The State and Nuclear Power: Conflict and Control in the Western World* (Seattle: University of Washington, 1984), 132.

Peace" policy, which literally liberalized trade in nuclear commodities, reflect what Gilinsky calls a "certain well-intentioned innocence in the notion that the best way to dissuade other nations from going after nuclear weapons was to provide them with everything nuclear but the bomb"? Was it in fact a "deliberate propaganda coup aimed at countering the USSR's offensive in the Third World"? Or did the industrial lobbyists finally achieve victory in Washington with the political acceptance of a policy which would "create a market for peaceful nuclear technology while the United States still had an undisputed lead in this field"?[28] There is some substance in all these explanations, perhaps even a coincident affinity amongst them. The point is that, when both the electricity and uranium-mining or private industries "live close to government", one can expect domestic and foreign policy to be affected, though the degree of affection exacted will remain an open question.[29] This will be increased when the state has an open investment in the nuclear reactor industry, as is the case in Canada, and when there is a strong link between a civilian nuclear industry and the defense establishment, as is the case with all the nuclear powers.

The general linkage between a certain sector of industrialists and the national security state is as prevalent in the southern hemisphere. One of the most vocal demands of industrializing elites, be they state managers, members of an aspiring entrepreneurial class, or international financiers considering significant investment, is for increased infrastructural development. Many of these groups, themselves part of a process of westernization that swept the decolonized world, were far from immune to the excitement the prospect of cheap nuclear energy generated in the North. Political elites have, meanwhile, found the prospects of mobilising nationalist sentiment enhanced by policies supporting what has been termed *nuclear nationalism* (Pakistan's Bhutto may be the most direct example). Having a research reactor is a matter of national pride, even if the long-term prospects for nuclear energy remain low in many southern states. The demand for nuclear weapons is no doubt related to the spiral

---

[28] Respectively: V. Gilinsky, "International Discipline Over the Uses of Nuclear Energy", in A. Wohlstetter, et al., ed., *Nuclear Policies: Fuel Without the Bomb* (Cambridge: Ballinger, 1978), 73-82, 74; Walker and Lonnroth, *op. cit,* 08; and M.A. Khan, "Nuclear Energy and International Cooperation: A Third World Perception of the Erosion of Confidence", in I. Smart, ed., *World Nuclear Energy: Toward a Bargain of Confidence* (Baltimore: Johns Hopkins University, 1982), 49-68, 52.

[29] The term is used by T. Price, in "Uranium Supply and Demand", in E. Meller, ed., *Internationalization: An Alternative to Nuclear Proliferation?* (Cambridge, Mass.: Oelgeschlager, 1980), 51-70, 69.

of insecurity in areas torn with regularized strife, in particular the Middle East and South Asia. But the developmental incentives, which make weapons proliferation and reactor mismanagement acute possibilities barring a functional regulatory regime, are as important.

The Northern and Southern demand for nuclear trade was the result of both push (from suppliers) and pull (from accepted assumptions about the correct path to development). This rise in demand was both anticipated and created by the major American firms (most notably, General Electric and Westinghouse) whose only significant competition in the industries' early days of American dominance was the Soviet Union, which had a limited (if captive) market. American dominance was apparent in technological terms, but was reflected in research and power reactor sales as well. Of the twenty-six nations which had reactors completed by 1963, the USA had supplied thirteen of them (the USSR six), and it supplied fuel to fifteen of these countries.[30] Once the decision to permit commercial trade was taken, nuclear power "was packaged and sold like most other commodities: in effect, free samples were provided; expensive product promotion campaigns were launched; and once the customers started appearing, the prices were raised."[31] There was, however, a strong difference: the need for extensive regulation, on some sort of international basis. In this light, early American dominance was temporarily solidified by the institutionalization of the Atoms for Peace regime with the creation of the International Atomic Energy Agency (IAEA) in 1957, which was to exercise a two-fold function, to promote peaceful uses of atomic energy while discouraging, by the implementation of a safeguards system, governments from pursuing military nuclear programmes.[32]

The Statute of the IAEA entered into force on 29 July 1957, after it was ratified by 18 states. It is headed by a Secretariat in Vienna, Austria, with a Director General (for the last four four-year terms this has been

---

[30] Walker and Lonnroth, *op. cit.*, 18. The legacy stands: the two largest Western competitors, the French Framatone and the German Kraftwerk, were created by Westinghouse and General Electric, respectively. See R. Finch, *Exporting Danger: A History of the Canadian Nuclear Energy Export Programme* (Montreal: Black Rose, 1986), 126. Indeed, the first U.S.-designed nuclear reactor export was a GE 16-megawatt boiling water reactor built in 1960 at Kahl, West Germany. P. Stoler, *Decline and Fall: The Ailing Nuclear Power Industry* (New York: Dodd, 1985), 38.

[31] R. Finch, *op. cit.*, 126.

[32] IAEA Statute, Art. II and III.A.5. See also L. Scheinman, *The International Atomic Energy Agency and World Nuclear Order* (Washington: Resources for the Future, 1987); and C. Jonsson and S. Bolin, "IAEA's Role in the International Politics of Atomic Energy" in L. Finkelstein, ed., *Politics in the United Nations System* (Durham: Duke University, 1988), 303-323.

Hans Blix), but the Executive is a 35 member Board of Governors, both designated and elected. It is funded by member states in accordance with a UN scale of assessment (Article XIV.D) and by voluntary contributions (Art. XIV.G). By 1992 the Agency had 114 members, 2135 staff members, 200 full-time inspectors in the Department of Safeguards, and a proposed budget for 1994 of US $201.1 million. Member States have also been asked to approve a target of US $58.5 million for voluntary contributions to the IAEA's Technical Assistance and Co-operation Fund. This fund helps cover IAEA programs designed to spread involvement in nuclear power production. For example, the purpose of the Irradiated Fuel Management Advisory Programme (IFMAP), which addresses various issues related to both power and research reactors, "is to offer guidance and training primarily to specialists in developing countries."[33] The Agency's Division of Nuclear Fuel Cycle and Waste Management is presently administering 42 technical co-operation graphs designed to assist countries in exploring and developing their uranium resources. Countries include Algeria, Argentina, Chile, China, South Korea, Egypt, Gabon, Indonesia, Iran, Madagascar, Malaysia, Mali, Mongolia, Pakistan, Philippines, Portugal, Syria, Thailand, Venezuela, Vietnam, and Zambia. This work includes environmental impact assessment, airborne radiometric and magnetic surveys (one covered more than 30 000 km^2 in the Central Belt Area of Peninsular Malaysia), natural radioactivity mapping, airborne gamma spectrometric surveys, and analytic laboratory and computer processing facilities. A coordinated research programme known as BEFAST will use data from industrialized states on spent fuel storage to help the future storage planning in developing countries. Evidently, the promotional role is still central to the operations of the IAEA, regardless of the publicity surrounding its recent involvement in weapons inspections in Iraq. In this sense the nonproliferation regime predates the NPT, though that Treaty is now its integral feature, by over a decade.

4. The Nonproliferation Regime

From its inception, one of the normative (and unchallenged) pillars of the nonproliferation regime has been the belief that nuclear technology, when not applied to the creation of weapons of mass destruction,

---

[33] Semenov and Oi, *op. cit.*, 6. Semenov and Oi are IAEA officials. The following updated information of about IAEA activities is found the latest issue of the *IAEA Bulletin* (1993).

is beneficial to humankind, and the IAEA has a mandate to promote the spread of this technology. Because of widespread consensus that nuclear weapons, which may be produced in conjunction with civil technology and plant, pose a grave danger wherever they exist, exceptional diplomatic measures were necessary in order to legitimize nuclear commerce on an international level. One IAEA official writes: "several industry representatives have told us that without the Agency safeguards systems there would be little or no industry, and little or no nuclear trade."[34] What other industry is so dependent upon international collaboration in the diplomatic arena itself?

That governments are deeply involved is hardly surprising. Most states with an advanced nuclear industry have invested in exportation, by extending credit and credit guarantees to importers and international banks, respectively. In the case of government subsidies by way of credit provision to customers, what takes place is, in essence, an "implicit transfer of money from the taxpayer, through the government, to the [exporting] company."[35] Camerilli, though quick to add that the "use of public funds to facilitate the export of nuclear goods and services was by no means unique to the U.S.", details the activities of Washington's Export-Import (Exim) Bank, which between 1958 and 1974 provided US $2.1 billion for the construction and fuelling of 42 exported nuclear power projects and, during the sales slump of 1974 to 1980, made over twice that amount available in direct loans and loan guarantees, mainly to South Korea, Taiwan and the Philippines.[36] Worseley cites the Westinghouse-Philippines-Eximbank deal as "the largest export order in the history of the nuclear industry."[37] Indeed, financing by state capital has been an important element in stimulating reactor sales, especially to developing countries. Loans on clearly concessionary terms were an important element in a United States sale to India and in a Canadian sale to Pakistan. Elsewhere, export financing arrangements, even though on nonconcessionary terms, were an important catalyst of sales.[38]

---

[34] M. Wilmhurst, "The Adequacy of IAEA Safeguards For the 1990s", in M. Fry, et. al., eds., *Nuclear Non-Proliferation and the Non-Proliferation Treaty* (Berlin: Springer-Verlag, 1990), 13-18, 15.

[35] J. Falk, *Global Fission: The Battle Over Nuclear Power* (New York: Oxford University, 1982), 78.

[36] J. Camilleri, *op. cit.*, 240.

[37] P. Worsely and K. Buenor Hadjor, eds., *On the Brink: Nuclear Proliferation and the Third World* (London: Third World Review, 1987), 36. The post-Marcos Philippines has since renounced nuclear power, but not without incurring the debt associated with the Marcos experiment.

[38] B. Goldschmidt and M. Kratzer, "Peaceful Nuclear Relations: A Study of the Creation and the Erosion of Confidence", in I. Smart, *op. cit.*, 19-48, 30.

It should be apparent from the above quotation that American firms would have to compete not only with overseas and Canadian counterparts, but with an onslaught of quite participatory foreign governments as well. It may be reasonably hypothesized that, in abandoning the initial policy of denial, Washington was working hard to maintain its monopoly -- by creating dependence upon its own exports -- while acknowledging the inevitability of nuclear development elsewhere. Ironically, the initial denial policy, which came into effect in 1946 as the U.S. Atomic Energy Act cut off "even the flow of information on civil nuclear power development which has been anticipated under wartime agreements" may have spurred nuclear development in Europe.[39] The British and French, in particular, were inclined to see the Act as an "American slap in the face [which] left a legacy of deep bitterness."[40] There was a certain paternal demeanour about the manner in which American political elites approached the issue, illustrated by the ill-fated "universalism" of the doomed Baruch plan, and evident within the North-South context as well as early American policy towards other (competing) industrialized nations:

> U.S. supremacy in strategic nuclear affairs was largely accepted within the Western Alliance. But its early monopoly in civil nuclear production was resented, and regarded by other industrial nations as a threat to industrial welfare. Escaping that monopoly became the object, at corporate and governmental levels, of vigorous policies of industrial and technological development in Europe and elsewhere.[41]

Declining American dominance in reactor sales and related technology may be seen, then, as the result of an "inevitable maturation process in the reactor industries of other advanced industrial states."[42] Several nations, notably Canada, West Germany, France, and the U.K., became involved with exports during the late 1960s and 1970s. A drastic slowdown in the domestic nuclear market coincided with threats to the

---

[39] A. Kapur, "Nuclear Energy, Nuclear Proliferation and National Security: Views From the South", in J. Keeley and R. Boardman, eds., *Nuclear Exports and World Politics: Policy and Regime* (New York: St. Martin's, 1983), 163-193, 164-165.

[40] W. Patterson, *op. cit.*, 04.

[41] Walker and Lonnroth, *op. cit.*, 18.

[42] E. Wonder, "The Nuclear Marketplace and the Nonproliferation Regime", in J. Pilat, ed., *The Nonproliferation Predicament* (New Brunswick, N.J.: Transaction Books, 1985), 71-82, 73.

nonproliferation regime such as the Indian implosion of a plutonium device in 1974, questionable deals between European suppliers and Latin American customers, reported weapons' development (and near-use) in Israel, South Africa, and Pakistan, the continued escalation of the superpower arms race, and the beginning of apprehension about what are now commonly referred to as the emerging nuclear suppliers. The list of the latter is lengthy, including Argentina, Brazil, India, Israel, Japan, Pakistan, China, South Africa, South Korea, Spain, and Taiwan. These challenging events and developments all served to shake the vulnerable "Calderesque balance"[43] which had been established by American and Soviet diplomacy, and which operated in a leadership vacuum in the post-American dominance era. But the loss of technological monopoly may have had just as much to do with a loss of confidence in American international leadership, despite the fact that the U.S. remains the principal financial contributor to the IAEA. Influence within that organisation has assumed a new, diffused character, expressed by a recent Ambassador to the IAEA, Richard Kennedy:

> We no longer possess the degree of influence in the nuclear field -- scientific or commercial -- that we once enjoyed ... As mastery of the technology has become more widespread, the ability to influence others through a nuclear supply relationship, let alone dictate their nuclear energy choices, has diminished.[44]

American-inspired exercises in management control, such as the 1977 International Nuclear Fuel Cycle Evaluation (INFCE) under the Carter Administration, and recent hard-line reactions to Iraqi and North Korean intransigence, have not altered this reality.

[43] The metaphor, which refers to the mobiles created by the American artist Alexander Calder, is used by Bujon de l'Estang. The nonproliferation regime "is a finely-designed mechanism of many quite diverse elements, all tied together by a thin wire ... if you disturb an element here, it jiggles over there -- and it may even collapse." *Op. cit.*, 131. (Such notions of balance may be applied to arguments against nuclear power itself, since ecology is the ultimate Calderesqueian structure.)

[44] R. Kennedy, "Nuclear Trade: Reliable Supply and Mutual Obligation", in R. Jones, et. al., ed., *The Nuclear Suppliers and Nonproliferation: International Policy Choices* (Lexington: Lexington Books, 1985), 30.

5. Contemporary Nuclear Trade Politics

Despite this decline in American dominance, however, the non-proliferation regime has remained essentially intact: though several potential suppliers have not signed the NPT, Goldblat and others believe "the fact that the NPT has managed to attract more adherents than any other arms control agreement clearly carries political weight"[45] and this is primarily because of the explicitly universalistic appeal of the Treaty. But suppliers themselves -- despite the trigger lists of the Zangger Group and London Club -- have yet to reach the type of export policy consensus needed for anything approaching an airtight system of control, even if an effective monitoring regime was possible.

Figure 4.1: Global Nuclear Power, 1992

Country	Units[a]	Country	Units[a]
Argentina	3	Korea, rep. of	9
Belgium	7	Lithuania	3
Brazil	2	Mexico	2
Bulgaria	6	Netherlands	2
Canada	22	Pakistan	1
China	3	Romania	5
Cuba	2	Russian Federation	46
Czech Republic	6	South Africa	2
Finland	4	Slovak Republic	8
France	61	Slovenia	1
Germany	21	Spain	9
Hungary	4	Sweden	12
India	14	Switzerland	5
Iran	2	Taiwan	6
Japan	53	United Kingdom	38
Kazakhstan	1	Ukraine	21
		United States	112

a: includes plants under construction

[45] J. Goldblat, "The Nuclear Non-Proliferation Imperative", *Bulletin of Peace Proposals*, 17: 3-4 (1986), 491-496, 491. Indeed, NPT membership is expanding. In 1992, the Treaty was joined by Estonia, Latvia, Slovenia, Uzbekistan, Azerbaijan, Namibia, Niger, Myanmar (Burma), and, most importantly, France and China.

If Nuclear Suppliers Group (NSG) guidelines on safeguards (and the IAEA's implementation of them) are intended to constrain the dissemination of technology assumed *a priori* to be dangerous (because of the possibility of weapons production or, perhaps, lack of environmental management in its use), this implies, at the least, "a supplier oversight function, a judgement of the 'legitimate' needs of nuclear recipients."[46] Above the obvious discrimination this creates in terms of North-South relations, we might ask whether such a harmonic policy approach is indeed possible in a competitive trade environment on the supply side.

Figure 4.2: Nuclear Power as a Percentage of Energy Generation, Selected Countries, 1992

Country	Percentage	Country	Percentage
Lithuania	80	Japan	27.7
France	72.9	Ukraine	25
Belgium	59.9	United Kingdom	23.2
Slovak Republic	49.5	United States	22.3
Hungary	46.4	Czech Republic	20.7
Rep. of Korea	43.2	Canada	15.2
Sweden	43.2	Argentina	14.4
Switzerland	39.6	Russia	11.8
Spain	36.4	South Africa	6
Taiwan	35.4	Netherlands	4.9
Slovenia	34.6	India	3.3
Finland	33.2	Mexico	3.2
Bulgaria	32.5	Pakistan	1.2
Germany	30.1	Brazil	0.7

Source for both figures: *IAEA Bulletin*, 3/1993, page 52.

Such harmony may be imagined in a world of sovereign states that each perceive a mutual interest in striving for and adhering to a strict trade regime. This oversimplification fails to convey either the nature of the contemporary international system or the fantastic complexity of the proliferation issue-area. The importance of transnational factors in world

[46] J. Keeley, *op. cit.*, 103.

politics is demonstrated by an even cursory look at nuclear trade, on both supply and recipient sides; this is especially so in the North-South context. We are dealing with a transnational phenomenon on two levels, one empirical and one normative, both affected by the decline of American dominance described above. No nonproliferation movement, in particular one aimed at safeguarding environmental security by rejecting the spread of both nuclear weapons and negligent nuclear power operations, will be successful if concentration is directed solely at the state level of interaction.

International commerce of any type involves a plethora of actors whose behavior is at least partly attributable to self-interest. One of the central dynamics within the nonproliferation regime takes place at the state-capital nexus: exporters pursue profit within a context of domestically and externally imposed regulations, and there is no doubt that some have been more successful than others in escaping these trade confines. Nuclear merchandise and technology may not be good business all the time, but it is always big business. The major firms involved in the world market, such as Babcock and Wilcox, GE, and Westinghouse from the U.S., Kraftwerk Union from Germany, Mitsubishi and Toshiba from Japan, Nuclear Power Corp. from the U.K., Framatome from France, and Atomic Energy of Canada[47] are powerful and often highly diversified corporations wielding considerable influence on policy. The energy sector, traditionally centralized, has some autonomy in civil affairs,[48] though nuclear power has proven to be an issue of such extraordinary public consternation there are limits to this autonomy.

The safeguard leniency which caused such concern in the 1970's may be seen as an example of state-industry policy convergence. France and (what was previously) West Germany have typically been isolated in this regard for permitting extensive transactions with non-NPT, "threshold" states, in particular South Africa and Brazil, respectively. The

---

[47] We might note that only the Canadian example is wholly state-owned, and it incorporates a variety of private firms into overall structure. Atomic Energy of Canada (which hasn't sold a reactor since 1978) is currently aiming at Asian markets with Candu reactors larger than 700 megawatts, even though the pro-nuclear lobby in Canada is now pushing for the development of smaller reactors for domestic consumption. See Sheila Malcolmson's Letter to the Editor, *The Globe and Mail*, July 28, 1993.

[48] As Falk (*op. cit.*, 332) puts it, "ownership of technologies is a major factor in constraining the degree to which people can participate in the decisions made about them." The nuclear power and arms race issues have proven fertile ground for public resistance movements in both western and even eastern political cultures at least partly because of this non-participatory political aspect.

German-Brazil deal was, according to Spence, "intended to give West Germany secure access to uranium supplies in the face of increasing friction with its traditional supplier, the USA."[49] Spence adds that the U.S. firms had not been able to close the deal with Brazil because American policy requires full-scope safeguards. A similar situation arose with regard to Argentina, which "bought a German nuclear-power plant and a Swiss heavy-water plant at far higher cost than a similar plant offered by Canada [because] Canada insisted on full fuel-cycle safeguards as a condition of supply [and] the FRG and Switzerland did not."[50] The new emphasis in the proliferation literature on the emerging suppliers is due to a belief that, "in general, it appears that there may be less policy incompatibility (between private firms and state bureaucracies) regarding nuclear exports among most than is the case for the traditional suppliers."[51]

Again, nuclear trade takes place within the context of a competitive international economy. Industries, faced with underconsumption in home markets, may be driven by the demands of domestic competition and opposition on environmental grounds to seek exterior markets in order to justify heavy long-term investments (and many governments have also made similar investments). There is an irony to the environmental politics of nuclear power, since successful opposition in the west has helped contribute to its spread world wide; a parallel with both industrial and agricultural pollution may be drawn. As firms compete, their willingness to engage in non-regime behavior may well increase as it becomes necessary to dispose of surplus production. Without an unchallenged monopoly held by any one or related set of firms, as has been the case since the late 1960's, it becomes increasingly difficult to ensure regime standards are respected. One of the more fundamental threats to any supply regime will be its inherent corporate and national competition.

We might qualify the above statement somewhat. Domestic underconsumption is not the only trade stimulant. France remains the sole traditional supplier state which relies primarily upon nuclear power for

---

[49] M. Spence, "Exploring the 'Peaceful Atom': The Pattern of Nuclear Proliferation", *Radical Science* 14 (1984), 80-93. Of the French-South African deal of 1976, Camilleri writes: "France's readiness to offer favourable credit terms and adopt a more relaxed attitude towards safeguards' requirements had been sufficient to clinch the deal" (*op. cit.*, 238); see also Z. Cervenka and B. Rogers, *The Nuclear Axis: Secret Collaboration Between West Germany and South Africa* (London: J. Friedman, 1978).

[50] D. Fischer, "The Role of the IAEA", in J. Simpson and A,. McGraw, eds., *op. cit.*, 55-68, 60.

[51] E. Solingen, "Brazil: Technology, Countertrade, and Nuclear Exports", in W. Potter, ed., *International Nuclear trade and Nonproliferation: the Challenge of the Emerging Suppliers* (Lexington, Mass.: Lexington Books, 1990), 111-152, 140.

energy consumption, a circumstance which has strengthened its firm's competitiveness abroad, since "the volume and standardisation of reactor manufacture France has achieved at home have allowed it to bring production costs down below those of its competitors."[52] There have also been joint ventures, such as the GE and Hyundai project in South Korea. The arena of nuclear trade exists within an increasingly open world economy, a trend likely to continue in a post-Cold War era characterized by a turn towards pragmatism in foreign policy. In this light, Solingen remarks, the "1986 Argentine-Cuban agreement for nuclear cooperation and other Brazilian-Cuban negotiations point to [Latin America's] decreased emphasis on ideological barriers to economic and technological cooperation. The possibility of joint (or unilateral) Brazilian-Argentine nuclear exports to Cuba is more easily conceivable ... than ever before."[53] (Of course, the Americans would have something to say about this.) Though we are far from laissez-faire conditions, especially after the Iraq scare, there may be strong trends in that direction in the immediate future, trends the U.S. would find very uncomfortable and be relatively powerless to control.

The portent remains: the "newly industrializing countries", such as Taiwan and Brazil, which have managed to some extent to develop indigenous nuclear capability, are new competitors on the supply scene, and they will face dilemmas similar to those faced by Northern firms eventually. Potential buyers of nuclear commodities -- be they private utility companies, parastatal or state-managed infrastructural corporations -- will have a wider array of purchase options than could be imagined during the period of American dominance. With this, in cases where this is a possibility, will come a greater ability to circumvent safeguards and proceed towards nuclear weapons development. In a world system composed of sovereign states, complete transparency for the purpose of Treaty verification and IAEA monitorship is hardly possible, and the link between nuclear weapons and plutonium production remains. Two factors will delay a nuclearized world: the prohibitive costs (both financial and in terms of citizen opprobrium) of nuclear power, which still puts it out of reach for most Third World states; and an increasing concern with the environmental consequences of development, articulated in calls for sustainable development and soft energy technology and accepted, at least rhetorically, by major multilateral aid institutions today. As we will

---

[52] Walker and Lonnroth, *op. cit.*, 96.

[53] E. Solingen, *op. cit.*, 138.

see next, however, this latter factor may be overshadowed by more conventional normative concerns.

There are so many actors involved in international nuclear commerce that their aggregation is foolhardy. There are the industrial and financial players; scientific communities involved in research, design, and legitimation; military and security elites; employees of international organizations like the IAEA and Euratom; and the diplomats participating in NPT and Tlatelolco negotiations. What links these diverse actors are normative similarities which make the proliferation of nuclear power and, in isolated cases, weapons, feasible prospects for the southern hemisphere.

Perhaps the most luminous of all is the conviction that technological innovations are potential answers to social problems. As imperialist powers conquered what was subsequently known as the Third World, they brought with them modes of production that altered or even discontinued traditional lifestyles. Following decolonization, a tendency to look for solutions which embraced the Northern (be that capitalist or Stalinist) path of development has been reinforced by external aid. This quest for "modernization" has been spurred on by the structural exigencies of a world economy dominated by Northern capitalist states in which what Falk terms an "ideology of technical optimism"[54] prevails. Faith in 'big science' -- and nuclear physics is often viewed as the biggest, a perception begun with the mammoth Manhattan Project itself -- justifies what may in the long run prove to be hazardous investments of money and time in nuclear power. Power generation in the Third World has had a miserable record in general, from the effects of large hydroelectric dams to heavy dependence on coal use. Calls for the adoption of appropriate technology may help reduce the biospheric threats raised by this activity in the future; especially if the north is able to not only assist the south but to change its own ways.

The issue of technology and technological access is, of course, central to this discussion. One of the results of the plutonium fears of the 1970s with the widespread introduction of fast breeder technology was a new era of technological denial. But this cannot erase the fact that if southern political and economic elites feel obliged or determined to compete in the world capitalist system, they are in obvious need of technology to do so. Much of the diplomatic debate in the regime has centred around this idea: "having sold the world on the values of Western industrial

---

[54] J. Falk, *op. cit.*, 19.

know-how, the rich countries now face demands that they share expertise through mechanisms other than the marketplace."[55]

It has become a familiar complaint that, especially following the introduction of NSG guidelines, "nonparties [to the NPT] were able to obtain nuclear material, equipment and technology more readily and under less stringent conditions than NPT parties."[56] However, as Schiff suggests, the nuclear power debate may not, in all cases, be the real agenda of the Group of 77 (founded in 1964):

> Whereas the initial enthusiasm generated by the Atoms for Peace program was substantial, employment of nuclear power in developing countries has been slow. It is possible that the enthusiasm that now exists is partially a public stance on the part of developing country authorities seeking to commit developed countries to technical assistance in any possible area.[57]

Costly nuclear power is hardly a feasible energy alternative in the Third World today, but this rationality has become hostage to the competitive North-South context of the world system. Little room is left for thinking which gives conservation and the indigenous development of alternative energy sources (solar, wind, thermal, biogas, tidal, others) priority. The chairman of the Pakistan Atomic Energy Commission claims Third World elites are "allergic to any attempted overselling of these 'soft' technologies to them and regard such moves as yet another way of persuading them to give up their interest in nuclear power programs."[58] It is, however, quite possible that rethinking development economics along "small is beautiful" lines, in terms of more sophisticated cost/benefit analyses and sustainability, may have an impact.[59] However, if nuclear exports become a great deal cheaper or safeguard-free, this might not be the case, and the south will remain a prime target for future industry ex-

---

[55] B. Schiff, *op. cit.*, 163-164.

[56] J. Pilat and R. Pendley, ed., *Beyond 1995: The Future of the NPT Regime* (New York: Plenum, 1990), 04. This is debatable: it may be speculated that North Korea, for example, joined the NPT regime in 1985 precisely because North Korean officials believed it would increase access to nuclear technology.

[57] Schiff, *op. cit.*, 31 n19.

[58] M. Khan, *op. cit.*, 64.

[59] For a recent and well written treatment see Raymond Mikesell, *Economic Development and the Environment: A Comparison of Sustainable Development with Conventional Development Economics* (London: Mansell, 1992). I have reviewed this book in the December, 1993 edition of *Futures: The Journal of Forecasting, Planning and Policy.*

pansion. Eastern Europe and the former Soviet Union face a worse situation: they cannot afford not to keep operating dangerously inept nuclear power systems. The hope here lies in shifting to other forms of power or, at least, improving the management and technology of that which exists.

Parallel to the effort to industrialize is the attempt by security elites to have at their disposal the most sophisticated weapons available. There are determined 'nuclear nationalists' who have decided for themselves that nuclear power or even weapons are the sufficient technological fix. We may in fact discern a most disturbing (if understandable from a narrow strategic viewpoint) trend towards a quest for military modernity. This perception will only increase as a result of the 1991 U.S.- Iraq war, arguably the greatest demonstration of high-technology warfare ever. On this general theme, Ray writes of a "new generation of Western-style 'national security' elites who think and talk in similar security language to their Western counterparts. This contrasts with the earlier generation of nationalist elites, who were steeped in the political culture of post-colonial nationalism."[60] The perceived need for weapons of mass destruction has obvious implications for proliferation, but denunciations of such perceptions by nuclear powers smack of hypocrisy. This is one reason why it is absolutely vital that the Clinton Administration make good on its pledge to refrain from nuclear testing, and that other nuclear powers follow suit.[61] China's recent (October 1993) test does not make this impossible. Despite the disorderly nature of the post-Cold War world, there is a clear window of opportunity here. We have also recently witnessed the "first instance of nuclear rollback, a state which has unilaterally and voluntarily relinquished nuclear weapons", South Africa.[62]

---

[60] A. Ray, "Third World Perspectives on Security and Nuclear Non-Proliferation: An Indian View", in J. Simpson, ed., *Nuclear Non-Proliferation: An Agenda For the 1990s* (Cambridge, N.Y.: Cambridge University, 1987), 52-63, 62.

[61] It is also vital for sheer diplomatic reasons, since many nations have explicitly tied post-1995 NPT acceptance to the signing of a fully Comprehensive Test Ban Treaty. This is of course a contentious issue. See W. Epstein, "The Linkage Between a Nuclear Test Ban and Nuclear Non-Proliferation", in Fry, et. al., eds., *op. cit.*, 132-146, and V. Goldanskii and V. Davydov, "A CTB and the Prevention of Horizontal Nuclear Proliferation", 147-160 in the same volume; and L. Spector and J. Smith, "Deadlock Damages Nonproliferation", *Bulletin of the Atomic Scientists*, December 1990, 39-44.

[62] J.W. de Villiers, R. Fardine, and M. Reiss, "Why South Africa Gave Up the Bomb", *Foreign Affairs*, 72:5 (1993), 98-109. South Africa joined the NPT in July of 1991, concluded a safeguards agreement with the IAEA two months later, and presented it with an inventory of nuclear materials and facilities in October 1991. It has since been allowed to take a seat in the IAEA General Conference, from which it was previously barred by apartheid. Pretoria has accepted membership in the Zangger Committee (which lists nuclear exports that require

6. The Future of the Regime

The spread and indigenous development of nuclear technology, encouraged by the transnational factors discussed above, has changed forever the potential future directions atomic diplomacy may effectively pursue. The political economy of nuclear trade is dynamic, susceptible to market and regulatory factors. There have been many creative suggestions, from differing perspectives, concerning the NPT's extension conference of 1995.[63] Some authors are prepared to reject the NPT entirely while others defend it almost religiously.[64] Leaving aside the important CTB-linkage mentioned above, we may make several salient points at this time.

Ashok Kapur believes that the nonproliferation regime has been eroding for what are essentially security-related reasons: "first, because the regime lacks an enforcement mechanism against non-NPT states; second, because the institutions and legal mechanisms of the NPT regime are weak and ridden with loopholes; and third, because the proliferators are upwardly mobile in the international system and see themselves as regional powers."[65] Unlike Kapur, the present author believes the preservation of the Treaty is something worth striving for, and hopes that it will survive the upcoming extension conference[66] relatively intact (if not, regime instauration may not be possible). However, Kapur's realistic observations are all evidence that the NPT itself is not going to stop proliferation. Even as brief an overview as that presented in this chapter demonstrates how complex the political economy of nuclear trade is, especially in the post-American dominance phase. Indeed, the NPT sys-

---

safeguards); and it has applied for membership to the Nuclear Suppliers Group and the Missile Technology Control Regime.

[63] For a discussion of preparation for the Conference see Ben Sanders, "Non-Proliferation Diplomacy: Preparing for 1995", *Security Dialogue*, 24:2 (1993), 221-231.

[64] See for examples A. Kapur, "Dump the Treaty", *Bulletin of Atomic Scientists*, July/August 1990, 21-22; and B. Sanders, "Avoiding the Worst of All Possible Worlds", in J. Pilat and R. Pendley, eds., *op. cit.*, 17-26. Kapur has an article in the latter volume as well, "World and Regional Power Relations Without the NPT", 117-131. For a cogent if limited discussion see B. Roberts, "From Nonproliferation to Antiproliferation", *International Security*, 18:1 (1993), 139-173.

[65] Kapur, *op. cit.* in Pilat and Pendley, 118.

[66] That a specific date was set aside to review the Treaty in such a fateful fashion is, perhaps, a mixed blessing; while it challenges the Treaty, it forces world reflection upon the issue. See *The Treaty on the Non-Proliferation of Nuclear Weapons*, Article X (2). Though an NPT based on perpetual adherence may be a good thing, it is both unlikely and unnecessary, and periodic reviews, though agonizing, are helpful in allowing new factors (such as growing global environmental awareness) to enter into discourse.

tem was "overloaded with expectation since its inauguration"[67] and, as argued above, has served to legitimize an industry which, from the viewpoint of environmental security, has presented more of a threat than a benefit to humankind. Yet the issue-area, and the forum the NPT creates, presents an opportunity for increased international co-operation and, to anti-nuclear environmentalists, dissent.

There is a more general move towards making plutonium and highly enriched uranium stockpiles and production sites the regulatory target of an expanded regime.[68] This would allow for the creation of international norms regarding such obvious environmental risks such as large plutonium shipments from France to Japan, for the cleanup of radioactive areas of the former Soviet Union, and for nuclear waste disposal policies. It could also demand a high level of reactor safety in any Third World installations, irrespective of the particular export controls in place at the time. The obvious problem here will be an old one: such a regime would represent an intrusion on sovereignty, and would have to avoid the discrimination developing states have accused the NPT of perpetuating. But we should keep in mind our loose definition of a regime (and, judging by the operational problems of international organizations and international law, this should not be difficult). A loose framework, based on a plethora of multilateral organizations falling under the jurisdiction of several international organizations, seems the most one can hope for now. It might be possible, with international consensus, to discourage plutonium shipments on the oceans, part of the global commons. Even this is unlikely in the immediate future, given the relative weight of France and Japan in the international system. But it will be far more difficult, despite current American policy towards plutonium, to get nation-states of either hemisphere to put part of their energy planning under effective international surveillance and control, and clandestine interactions will not be stopped by formal diplomatic means.[69] This extends of course to include

---

[67] P. Power, "The Mixed State of Non-Proliferation: The NPT Review Conference and Beyond", *International Affairs*, 62:3 (1986), 477-491.

[68] See D. Albright, F. Berkhout, and W. Walker, *World Inventory of Plutonium and Highly Enriched Uranium 1992* (SIPRI: Oxford, 1993).

[69] The U.S. abandoned plutonium production for explosive purposes in 1988, and currently the Clinton Administration discourages even the civil use of plutonium. White House Fact Sheet on Non-Proliferation and Export Control Policy, Presidential Decision Directive 13 (PDD-13), September 27. Clinton reiterated this position in a speech to the UN General Assembly. See *Arms Control* 23:9 (1993), 27. The American policy will not oppose plutonium production by Japan or western Europe, "while it will seek to discourage such activity in regions of instability and high prospective risk." North Korea has come to dominate the

sales which involve former superpowers: the recent U.S.- Russia deal to ship enriched uranium to the United States for use in reactors after the dilution of weapons-grade enriched uranium does not fall under IAEA safeguards.[70] And such regimes have limits in terms of participation. For example, delegates to the November, 1993 London Convention conference voted for a complete ban on the dumping of radioactive wastes at sea, but Belgium, Britain, China, France, and Russia abstained.

Indeed, the nuclear fuel cycle necessarily involves many steps which threaten environmental security, from uranium mining and transport to reactor operation to waste storage, and a truly global nuclear regime would have influence in all these areas, domestic and export-oriented. The IAEA, as mentioned above, plays an active role in all steps of the fuel cycle. This involves both promotional and risk aversion activity. This is quite unlike the Organization for the Prohibition of Chemical Weapons (OPCW), established by the recently signed Chemical Weapons Convention. The OPCW "is responsible for achieving a complete ban on chemical weapons and is not responsible, at least as currently envisaged, for the promotion of peaceful uses of chemistry and chemical sciences."[71] These two verification mechanisms have much in common, including similar organizational structures, but the question of their collusion with industry separates them substantially.

Such a functional shift in the IAEA is unlikely, since it is part of the normative context discussed above. There are agreements in force that strengthen the risk-aversion aspects of the nuclear industry. Since 1974 the IAEA has operated its Nuclear Safety Standards Programme, complete with OSART (Operational Safety Review Teams) and ASSET (Analysis of Safety-Significant Events Teams). It was not until after the Chernobyl calamity, however, that the IAEA was "first confronted with the RBMK-type reactor."[72] The IAEA has attempted to harmonise notifi-

American nonproliferation agenda. J. Goldemberg, "Nonproliferation Nonsense", *New York Times*, Nov. 18, 1993, A19.

[70] K. Subrahmanyan, "Export Controls and the North-South Controversy", *The Washington Quarterly*, 16:2 (1993), 135-144, 138. See also O. Bukharin, "Soft Landing for Bomb Uranium", *The Bulletin of Atomic Scientists*, 49:7 (1993), 44-49.

[71] W. Dorn and A. Rolya, "The Organization for the Prohibition of Chemical Weapons and the IAEA: A Comparative Overview", *IAEA Bulletin*, 35:3 (1993), 44-48.

[72] M. List, "the West and Eastern Nuclear Power Plant Safety", *Aussenpolitik*, 44:4 (1993), 344-353. The IAEA, notes List, accepted the official Soviet account of the accident at the August 1986 Conference it convened. List believes that "the concern that the inadequate safety of Eastern nuclear power plants might discredit Western nuclear energy programmes too was ... one of the main motivational forces for the post-Chernobyl activities of nuclear power plant operators." *Ibid.*, 348.

cation of accidents with the INES (International Nuclear Event Scale). In October of 1986 a Convention regarding the early notification of nuclear accidents was signed, followed in 1987 by a Convention concerning assistance in the case of nuclear accidents. And in May of 1989 WANO, the World Association of Nuclear Operators, was formed, including all the then-nuclear power nations. This international organization takes familiar lines, since it serves to justify the continuation of the industry itself. It may even be necessary, in the future, to give the IAEA increased powers in order to centralize the environmental impact assessment and control of nuclear material.

But we are a long way off from such a widespread regime, with its attendant infringement on sovereign jurisdiction and demands for a mammoth international regulatory agency (the IAEA is clearly not up to such a task, though it will probably devote much more attention to safety issues in the future, if its international conference on nuclear power safety in September of 1991 is any indication). No one can deny the importance of monitoring nuclear product exports and (especially so in the North-South context) possible nuclear waste transport, nor of establishing early warning systems in the case of radiological emergencies. But we can question the very desirability of an institution with sweeping nuclear regulatory powers, however well they could possibly be enforced, from at least three perspectives. It would tend to represent the interests of its main financiers, the West and in particular advocates of nuclear trade. It would drain resources from possible international cooperation devoted to other energy paths. And it would serve to legitimize nuclear power itself, with the political power discrepancies associated with such energy centralization, the environmental risks, and the omnipresent threat of weapons diversion.

7. Conclusion

Ultimately, the assumptions upon which the pro-nuclear power nonproliferation regime are based must be challenged. However, it is highly unlikely that such a challenge will find emanation from the current diplomatic discourse, which is more involved with immediate nuclear politics and posturing. Events with the gravity of Chernobyl have not changed this, but the UNCED Conference may have a more lasting impression.[73] At the same time, Chapter 22 of Agenda 21, entitled "Safe and Environmentally Sound Management of Radioactive Wastes", calls in part for states, "in cooperation with relevant international organizations, [to] provide assistance to developing countries to establish and/or strengthen radioactive waste management infrastructures, including legislation, organizations, trained manpower and facilities."

There is room, one may hope, for a substantial extensity of North-South dialogue to include alternative energy options along the lines called for by a sustainable development agenda; and technological assistance in these areas (along with progressively effective debt relief) will contribute as much to global security, in particular environmental security, as a renewed NPT. Despite some recent sales, such as the reactor sales from China to Ghana and Syria, the great anticipated demand in the south for nuclear technology has not materialized. Writes one group of energy experts: "The high capital costs of nuclear facilities relative to other generating sources have served to make it less attractive, as has public opposition to nuclear power. The Taiwanese utility, for example, which during the 1970s spoke of having 20-24 nuclear reactors operating by the year 2000, now plans to have a total of only 8 units in operation by the year 2000."[74] This being said, scrapping the NPT altogether is not in anyone's best interest. Nuclear weapons remain the greatest single threat to humankind's survival. This symbolic significance should not be lost on us, nor should the very real market dynamics discussed above. Provided nuclear disarmament and a CTB is possible, the NPT makes security sense. The larger question may be whether actively promoting atomic power does as well.

Though his specific recommendations do not find favour here, we may conclude with Butler that future negotiations "will need to produce a true fusion of means to solve the problems of both the atom and development."[75] The regime will continue to be plagued by problems -- what isn't in world politics? And the nuclear industry, old and new, must comply with a nonproliferation agenda; the losses it as a whole may suffer

---

[73] However, we should note that the nuclear industry will use calls to stem global warming, which rang loudly from the UNCED Conference at Rio, to advocate nuclear energy. We should be cautious in accepting trade-offs when it comes to environmental security (for example, nuclear waste for less carbon dioxide in the atmosphere, or less flooding by large dam construction). Weighing the associated risks is at best a dubious exercise, and precludes not only other energy and development alternatives but conservation itself.

[74] S. Myers, et. al., "Prospects for the Power Sector in Nine Developing Countries", *Energy Policy* 21:11 (1993), 1123-1132, 1126.

[75] R. Butler, "Beyond the NPT", in Pilat and Pendley, *op. cit.*, 65-71, 70.

are insignificant next to the environmental dangers of a further nuclearized world. It is clear, however, that the radical perceptual shifting of the normative elements discussed in this chapter will not be the result of industrial or state initiative. Change will be as much the result of efforts outside the market and, for that matter, from outside the present discourse which facilitates its existence.

CHAPTER FIVE

WHALES: THE NORMATIVE TRANSITION OF THE IWC[1]

"... beyond the image of the planet earth itself, the most poignant symbol of the world environmental movement has been the whale ... The survival of whales continues to be an object of international anxiety ... and perhaps better than any other issue illustrates the difficulty of reconciling the multiple conflicting interests and values of nations in the management and protection of the biosphere."

Lynton Caldwell, 1990[2]

1. INTRODUCTION

Few wild mammals have received as much human attention as the Great Whales. Unfortunately, much of this attention has involved killing them; but a substantial measure was devoted to stopping their slaughter and establishing safeguards for their future. That the latter effort has (with evident qualifications) proven successful is testament to both the symbolic power of whales and the determination of those people who together managed to save them from extinction. This effort, and the

[1] Two earlier versions of this paper should be acknowledged. The first was presented to the Canadian Political Science Association Annual Meeting, Prince Edward Island University, in June of 1992, and was entitled "International Organization and Normative Transition: The Death of Whaling and the Survival of the Whale." The second is "International Politics and the Protection of Great Whales", in *Environmental Politics* (London: Frank Cass), 2:2 (1993), 277-302.

[2] *International Environmental Policy: Emergence and Dimensions*, Second Edition (Durham: Duke, 1990), 286.

ensuing saga, took place within the broad realm of international politics. Given our open conceptualization of security, and given the importance, in both ecological and social terms, of the whaling issue, it is argued here that the tangled issue of whale survival constitutes a modern global security dilemma.

With the normative transition of the International Whaling Commission (IWC) from a "whalers club"[3] to an official agent of conservation, the whale slowly begins to symbolise not humanity's inability to respect the beauty of nature, but our ability to cooperate in order to preserve it. This transition is of interest not only to ecologists and wildlife conservationists, but to scholars of world politics concerned with international cooperation and change. The whaling issue has brought to the surface stark contrasts between various foreign policy positions and, some claim, deep cultural factors, as well as overt tension between not only an increasing array of involved states, but between state and non-state, or sovereign-free, actors as well. In turn, the resultant regrowth of whale populations presents a challenge to the whaling regime as the value-split between conservationist (i.e., saving whales now for future consumption) and preservationist (i.e., invariant opposition to commercial whaling) philosophies becomes axiomatic.[4] The protection of great whales is a complex tale of political change and, currently, uncertainty; and it is an important one, given the symbolic power of the whale in this age of global environmental insecurity.

2. Wildlife Conservation and International Organization

As mentioned in our earlier theoretical discussion, it has become fashionable to refer to "regime formation" as the process of international organization which facilitates cooperation in the natural resources issue-area. Regime analysts might point to the whaling regime with some satisfaction, as an example of what contractarian conservationist regimes can do. While this is not disputed in its entirety, this chapter will show that regime analysis itself cannot explain the unique political concatenation

[3] This phrase is common. J. Cherfas, *The Hunting of the Whale: A Tragedy That Must End* (London: Bodley Head, 1988).

[4] For a recent discussion see M. Peterson, "Whalers, Cetologists, Environmentalists, and the International Management of Whaling", *International Organization*, 46:1 (1992), 147-186.

which conspired to convince the remaining whaling nations to stop. In particular, we must be aware of the biological characteristics of whales, the vital role played by non-state actors and, related to both, the decline in both demand for whale products and the social acceptance of whaling as a legitimate activity. These factors, buttressed by American economic leadership, combined to promote the essential physical change which occurred within the IWC in the late 1970s and early 1980s: a definitive expansion in the membership of non-whaling nations in the Commission. The IWC offers us an historical model of achieving relative -- though certainly not universal -- normative consensus when facing a crisis situation, and a demonstration of how international organisations remain pivotal "fora for choosing innovations" where the "re-examination of purposes is brought about by knowledge-mediated decision-making dynamics."[5]

In line with the note of caution that should accompany any optimism in the field of international affairs, we should not exaggerate the potential of such organisations. It is indeed difficult to speculate, as Robert Boardman once did, that global conservationist organisations may be unwitting "embryonic institutions of representative government for [non-human] species of the planet."[6] This would require a remarkable transformation in our collective anthropomorphic conception of democracy, as well near consensus on such matters, a context which clearly eludes current grasp. Nonetheless, the discernible normative transition which has taken place gives even the more cynical of international relations scholars and environmentalists reason for optimism.[7]

Although conservationist efforts which involve participants from different societies have become increasingly visible as nature has become

---

[5] E. Haas, *When Knowledge Is Power: Three Models of Change in International Organisation* (Berkeley: University of California, 1990), 8.

[6] R. Boardman, *International Organisation and the Conservation of Nature* (Bloomington: Indiana University, 1981), 184.

[7] For an interesting, if perhaps overly hopeful, example of optimistic speculation based on past normative transition, see J. Ray, "The Abolition of Slavery and the End of International War", *International Organization*, 43:4 (1989), 406-439. Ray reminds us that values and norms do play an integral part in the analysis of international politics; despite their emphasis on power relations, both Marxists, or what we have termed radical structuralists, and state-centric or "realist" analysts accept this. Indeed, discussion of the normative dimension in foreign policy goes back to H. Morgenthau's treatment of "International Morality and World Public Opinion" as "Limitations of International Power" in his classic *Politics Among Nations: The Struggle for Power and Peace* (New York: Knopf, 1949); and continues with C. Beitz, et. al., *International Ethics: A Philosophy and Public Affairs Reader* (Princeton University Press, 1985); and R. Falk, ed. *The Promise of World Order: Essays in Normative International Relations* (Philadelphia: Temple University Press, 1987).

increasingly threatened, there is a long history of such political behavior, especially in the field of international environmental law.[8] Both intergovernmental and transnational conservationist organisations, usually European and more often than not concerned primarily with economic factors, have been with us throughout this century.[9] As early as 1886 a *Treaty Concerning the Regulation of Salmon Fishing in the Rhine River Basin* was signed by Germany, Luxembourg, the Netherlands and Switzerland. The first international agreement to conserve a marine mammal was in all likelihood the *Fur Seal Convention* of 1911, signed by Japan, Russia, Great Britain (for pre-independence Canada) and the United States.[10] Another groundbreaking agreement was the U.S.-Great Britain *Migratory Birds Act,* signed in 1918.

In the contemporary setting what Robert Boardman has called the "linchpin of the system"[11] of international conservationist organisation is the *International Union for the Conservation of Nature and Natural Resources* (IUCN), formed in 1948 in conjunction with the United Nations Education, Scientific and Cultural Organisation (UNESCO). The IUCN is a unique umbrella organisation which covers intergovernmental and transnational conservation activity, often working laterally with a plethora of other organisations -- both state multilateral and non-governmental in composition -- which have achieved global significance. At the state level, the *Convention on International Trade in Endangered Species of Wild Fauna and Flora* (CITES) is often heralded as a diplomatic success; it has resulted in

---

[8] For early examples, see L. Leornard, "The International Protection of Whales", *American Journal of International Law,* No. 35 (1941), 90; S. Hayden, *The International Protection of Wildlife: An Examination of Treaties and Other Agreements for the Preservation of Birds and Mammals* (New York, 1942); L. Caldwell, "International Conservation Efforts: Cooperation to Protect Endangered Species and Environments", in his *In Defense of the Earth: International Protection of the Biosphere* (Bloomington: Indiana University, 1972), 53-9O; and L. Teclaff, L. and A. Utton, ed., *International Environmental Law* (New York: Praeger, 1974).

[9] On the early development of the latter, within an international context, see R. Boardman, *op. cit.,* 26-30. Early organizations included the International Council for Bird Preservation (1909) and the International Congress for the Protection of Nature (1913). The latter had planned to discuss the whaling issue as early as 1914 but European nations became quite justifiably preoccupied with more urgent matters in that historic year.

[10] Unfortunately, the Fur Seal Convention, which banned all pelagic commercial seal hunting in the North Pacific, instigated renewed interest in whaling in that area. See K. Dorsey, "Putting a Ceiling on Sealing: Conservation and Cooperation in the International Arena, 1909-1911", *Environmental History Review,* 15:3 (1991), 27-46; and D. Ehrenfeld, D. (1972), *Conserving Life on Earth* (New York: Oxford University Press, 1972), 18.

[11] Boardman, *op. cit.*

the current -- and controversial -- international ivory trade ban of 1989.[12] At the non-governmental or transnational level, "sovereignty-free actors"[13] such as Friends of the Earth[14] have organised on an international basis, with varying success, to pressure governments and industry to adopt conservation-oriented policies. Such purposeful informal interaction demonstrates the contemporary significance of issue-areas which evade the often confining grasp of diplomatic exclusivity customarily, and jealously, attached to issues of "high politics" by state representatives. This is reflected in the academic community by a more general concern with what James Rosenau has termed the "transnationalisation of world affairs" or "the processes whereby international relations conducted by governments have been supplemented by relations among private individuals, groups and societies."[15] We have dealt with this issue at some length in our discussions of the changing nature of international security concerns, as well as the concept of environmental security itself.

Nonetheless, it is obvious that governments will remain the formal source of wildlife legislation and, one hopes, assume most of the responsibility for its subsequent enforcement. This is as true in the case of issues of the commons as it is of areas where domestic regulation itself will suffice. In the international case, governments direct energy and, some might argue, cede some sovereignty to intergovernmental organizations. The International Whaling Commission is the body with which we are concerned in this chapter. But, referring back to our broad perspective on regimes, and Young's distinction between organisations and institutions, it is not enough to study the IWC, which did not establish a permanent office until 1976 (in Cambridge, England), and remains a relatively small international organisation, itself. We must look at the institution of the contemporary whaling regime and the actors which forged it.

---

[12] See J. Macgregor, "The Paradoxes of Wildlife Conservation in Africa", *Africa Insight* 19: 4 (1989), 201-212; and M. Trexler, M. (1990), "The Convention on International Trade in Endangered Species of Wild Fauna and Flora: Political or Conservation Success?" Phd. Dissertation, University of California, 1990 (abstract on CD-Rom). J. S. Adams and T. McShane stress the vital human dimension of conservation efforts in their recent *The Myth of Wild Africa: Conservation Without Illusion* (New York: Norton, 1992).

[13] J. Rosenau adopts this term in *Turbulence in World Politics: A Theory of Change and Continuity* (Princeton: Princeton University Press, 1990).

[14] See in particular T. Burke, "Friends of the Earth and the Conservation of Resources", in P. Willetts, ed, *Pressure Groups in the Global System: The Transnational Relations of Issue-Oriented Non-Governmental Organisations* (London: Francis Pinter, 1982), 105-124.

[15] J. Rosenau, *The Study of Global Interdependence: Essays on the Transnationalization of World Affairs* (New York: Nichols, 1980), 1-2.

In recent years this institutional network has expanded, so that its members include -- at an influential level -- non-whalers. This has, it is argued, paved the way for an historic, though non-universal, normative transition which has in turn redefined the international organisation which served as the focal point for this transformation.

We have already mentioned that there is a limiting quality to any regime theory which draws rather arbitrary parameters around issue-areas. This caveat is vital if we seek an enhanced understanding of the international politics of wildlife conservation, since responding to the ominous threat of biological simplification or species extinction not only transcends state-centrism, but involves many complex overlapping temporal spaces of human-nature interaction. This complexity is frustrating, since human activity is without any doubt causing increased species extinction.[16] James McNeely, writing for the IUCNNR, asserts that the three major threats to biological diversity as a whole are habitat alteration (usually from highly diverse natural ecosystems to less diverse agro-systems), over-harvesting, and climate change.[17] If the gravest danger to wildlife is human encroachment upon and alteration of habitat, any attempt to preserve cetaceans must involve a parallel effort at preserving their fragile environments, the oceans. Such activity will inevitably require significant international cooperation amongst political and industrial elites and, importantly, greater participation by concerned citizens.

Furthermore, the plight of the Great Whales engendered much public attention because of their very natural and unique biological and behavioral characteristics, and this had less to do with immediate regime dynamics than with human perception and reaction. If the symbolic power of the Great Whale was partially responsible for generating preser-

---

[16] There has been a rapid growth of literature. A small sample should include L. Kaufman, "Why the Ark is Sinking", in same and K. Mallory, ed., *The Last Extinction* (Cambridge: MIT, 1986), 1-41; P. Ehrlich and A. Ehrlich, *Extinction: The Causes and Consequences of the Disappearance of Species* (New York: Random House, 1981); J. Swaney and P. Olson, "The Economics of Biodiversity: Lives and Lifestyles", *Journal of Economic Issues*, XXVI:1 (1992), 1-26; and D. Paterson, *The Deluge and the Ark: A Journey Into Primate Worlds* (Boston: Houghton Mifflin, 1989).

[17] See J. McNeely, et. el., (1990), *Conserving the World's Biological Diversity* (Washington: IUCN, 1990); and J. McNeely, *Economics and Biological Diversity: Developing and Using Economic Incentives to Conserve Biological Resources* (Switzerland: IUCN, 1988). Two analysts listed the four major threats to species in Europe, in order, as follows: 1)loss of habitat; 2)degeneration of environment; 3)disturbance of biological balance; 4)pursuit by man. That all four are interrelated is obvious. R. van der Woude and A. Van Wijngaarden, *Animals in Danger: A Study of Certain Mammal Species Threatened with Extinction in Europe* (Council of Europe, 1969); see also R. Boardman, *op. cit.*, 10.

vationist efforts on its behalf, perhaps this charisma can be utilised to re-orient human perspectives towards the negligent use of its vital habitat.

3. The International Politics of Whaling

In the whaling season of 1961/62, a peak kill of more than 66,900 large whales was registered with the International Whaling Commission.[18] In 1982, less than twenty years later, the IWC voted to ban all commercial whaling by 1986, and by 1988 this moratorium -- after the anticipated stalling by whaling nations ended -- was in full effect. Sadly, of course, catch statistics have dwindled rapidly between these dates simply due to overwhaling. It is also true, however, that in the intervening years a remarkable transformation has taken place at the political level.

Whaling seems inextricably linked to the early histories of most naval powers. It is said that the Basques were whaling before the tenth century; Norwegians were whaling as early as 1250; the Dutch in 1500, joined soon after the French and English; and the early American whaling trade is well popularized by literary classics such as "Moby Dick". Historical accounts of early commercial whaling are proliferating, and they reveal much about the relationship between industrial expansion, human-nature interaction, and international relations.[19] The history of commer-

[18] Bureau of International Whaling Statistics, *International Whaling Statistics* LXI, (Oslo, 1968), 8. This was achieved with an army of 48 shore stations, 26 "floating factories", and 444 catcherboats. Since the original drafting of this study, the Russian government has admitted that Soviet Union catch reports were falsified, which means almsot all figures (specially humpback and blue whale catches) are low. The Russian government does have the actual figures (apparently, the KGB doctored them for IWC purposes during the Cold War), and an IWC representative is calculating more accurate figures. The episode does throw in question any monitoring system not based on external observation. According to one cetologist, this information helps explain why whale populations have been so slow to recover from whaling. Interview with Ray Gambell, Executive Officer of the IWC, March 6, 1994.

[19] A short list would include R.L. Webb, *On the Northwest: Commercial Whaling in the Pacific Northwest, 1790-1967* (Vancouver: UBC, 1988); E. Stackpole, *Whales and Destiny: The Rivalry Between America, France, And Britain for Control of the Southern Whale Fishery, 1785-1825* (Amherst: University of Massachusetts, 1972); D. Francis, *Arctic Chase: A History of Whaling in Canada's North* (Toronto: Breakwater, 1984); J.R. Bockstoce, *Whales, Ice, and Men: The History of Whaling in the Western Arctic* (Seattle: University of Washington Press, 1986); G. Jackson, *The British Whaling Trade* (London: Adam and Charles Black, 1978); R. Harrison, "History of Whaling", in same and M. Bryden, eds., *Whales, Dolphins and Porpoises* (London: Merehurst, 1988), 182-195; A. Barrow, "The North-East Coast Whale Fishery, 1750-1850", Phd. Dissertation, Council for National Academic Awards (U.K.), 1989 (abstract on CD-ROM); C. Sanger, "The Origins of the Scottish Northern Whale Fishery", Phd. Dissertation, University of

cial whaling carries with it a certain mystique, from the brave early days of sail to the latter days of the mercilessly efficient factory ship.

Whaling was in effect one of the economic activities which facilitated the spread of European civilization into previously uncharted waters and areas, and it has evolved its own unique folklore and mythology. Whale products have contributed a good deal to the development of industrial society: whale blubber, once burnt on whaling vessels themselves, provided oil for lubrication, candles, and reading lamps. Indeed, America's early interest in whaling began to wane with the discovery of petroleum in Pennsylvania in 1859. Blubber was also used to make soap, and the explosive nitroglycerine was made from one of the byproducts of the soap production process. The list of whale byproducts, extended into the current century, is truly amazing; and, while whales were perceived as abundant, humankind was slow to develop necessary substitutes.[20] But if whaling contributed, in this indirect and commercial fashion, to the development of international society, it also often involved considerable international conflict over disputed whaling grounds (a treaty between Russia and the United States, prompted by a dispute over whalers' access to the Russian Pacific, was signed as early as 1824).

Whaling was at best a challenging financial enterprise, prone to recurring boom and bust cycles. These cycles partly reflected broader currents in the economies of whaling nations, but also reflected the human drive to see a quick return on initial investment on the human side, and the biological inability of 60-ton creatures to reproduce themselves like fish. As a result, whales were threatened with extinction even during the age of wind, before the quicker and more mobile steamships prowled the seas. The Atlantic Gray was wiped out along the coast of Europe around 500 A.D., and its North American counterpart in the early 1700s, while Basque whalers had overexploited Right Whales in the Bay of

---

Dundee, 1985 (abstract on CD-ROM); and what is probably the most widely read and authoritative, if outdated, *The History of Modern Whaling,* by J. Tonnessen, and A. Johnsen, trans. by R. Christophersen (Los Angeles: University of California Press, 1982).

[20] A partial list includes flexible baleen filaments for watch-springs and umbrellas, oil for a variety of purposes, including illumination, plastic, margarine, lipstick, detergents, brushes, engineering coolants, fertiliser (guano), medicine, gelatine, and corsets. Spermaceti oil, obtained from the Sperm Whale's *(Physeter macrocephalus)* head, has been used in space research; and ambergris -- a form of excrement found in the Sperm's intestine -- has been used as a fixative in the perfume and cosmetic industries. Today, the only really valuable whale product is meat, for limited human consumption and animal feed, and we will discuss this later.

Biscay by the 13th Century.[21] But there are both quantitative and qualitative differences between hunting by wind and sail and modern commercial whaling.

The evolution of modern pelagic whaling has been spurred by the advent of certain technological improvements in nautical prowess and killing and flensing efficiency: the harnessing of steam, the construction of shore-based whaling stations, the grenade harpoon, the inflation devices[22] which allowed carcasses to float, the "factory" ships introduced by the Norwegians in the 1920s. Of particular significance were the technological contributions of the Norwegian Svend Foyn, who in the mid-to-late 1800s developed, among other things, the exploding grenade harpoon, the first modern whaling station, a steam-winch for hauling carcasses bound for processing, and a system for making "guano" (fertilizer) from leftover bones.

Writes historian Robert Webb: "The efforts of the mythic Svend Foyn became the earliest legends of modern whaling as advancing industry resolved the herculean joust between man and whale into a unilateral cannon-bombardment."[23] Factory ships were of particular significance: several catcher boats could service one of these gigantic vessels, which made even the mammoth Blue Whale (*Balaenoptera musculus)* look small, and the Japanese and Soviets would base their emerging whaling industries on their use. Any pretensions concerning the romance of whaling were quite indubitably banished to the realm of fiction with these leaps in killing technology. There is a symbolic progression discernible: once a great struggle with nature, capturing whales became a methodological commercial function of applied technology.

---

[21] See J. Williams and R. Nowak, "Vanishing Species in Our Own Backyard: Extinct Fish and Wildlife of the United States and Canada", in L. Kaufman and K. Mallory, eds., *op. cit.*, 107-139, 109.

[22] Whales sink when killed, with the single exception of the Right Whales (*Balaena glacialis, balaena mysticetus,* and *caperea marginata),* which float. Indeed, this is whence its name originated; early whalers, aware of this peculiarity as well as its slow swimming speed and high yields of baleen and oil, determined it the "right" whale to kill.

[23] Webb, *op. cit.*, 142.

Figure 5.1: Selected Hunted Great Whales

Species	Characteristics	Status
Sperm	temperate\polar regions;	Slow recovery
	60-70 feet; 15-70 tons; lifespan:	
	70 yrs; 14-16 month gestation;	
	valued for spermaceti and meat.	
Blue	polar\tropical regions;	80-110; Endangered
	feet; up to 190 tons; lifespan:	
	80 yrs.Close to extinction,	
	valued for oil and meat.	
Fin	polar to tropical; over 80 feet;	Vulnerable
	70-80 tons.	
Sei	polar to tropical; 59 feet;	Not endangered
	30 tons.	
Brydes	tropical and subtropical seas;	Not endangered
	43 feet; 26 tons.	
Minke	polar to tropical; 36 feet; 10	Good recovery
	tons; strong demands for a	
	resumption of hunt for this	
	species, for meat only.	
Grey	Pacific coastal waters; 39-50	Slow recovery
	feet;16-34 tons; 13 month	
	gestation; life span: over 70 yrs.	
Right	temperate, both hemispheres;	Endangered
	16-60 feet; 50-56 tons; gestation:	
	10-11 months; life span: 30 yrs;	
	heavily hunted because they were	
	slow and floated when dead.	
Bowhead	arctic basin; 11-66 feet;	60-80 Endangered
	tons; gestation: 10-11 months;	
	lifespan: over 30 yrs.	
Humpback	polar to tropical; 50-60 feet;	Endangered
	40-65 tons; life span: 95 yrs.	

Sources: *Sea Mammals* (New York: Torstar, 1984); S. Savage, *Endangered Species: Dolphins and Whales* (Hong Kong: New Burlington Books, 1991). We should note that little is known about most whale species, including accurate assessments of current populations.

It is a popular error to consider the near-extinction of Great Whales a true "tragedy of the commons." Rather, the tale is one of unmitigated greed on the part of several actors which, supported by the state apparatus of the sovereign nations in which they chiefly operated, went about whaling with no regard for the future of whales or, for that matter, whaling. As G. Kent writes, the "excessive whaling by Japan and the Soviet Union ... represents the exercise of oligopolistic power more than it represents a tragedy of unlimited access by large numbers of equal competitors."[24] The tradition of excess was, in fact, begun in the 1800s. Though it may have been mistakenly believed that whales were capable of rapid multiplication, even slightly experienced whalers could not help but know better. As Robert Webb writes, "[i]t is a parable of the whale fishery, now well understood, that each new whaling area yields up the bulk of its harvest in the first few years of whaling. Increased use only depletes the natural resources and leads to a failure of the ground. This proved to be as true in the western Arctic as it had been on the [Pacific] Northwest Coast. The Arctic whalemen hunted further and longer and, eventually, the bowheads were extirpated like their more southerly cousins."[25] It is not difficult to identify or recognise near-extinction from overhunting of most species, and the depletion of large cetaceans -- which must be caught individually -- is particularly visible.[26] Tellingly, during the brief respite from whaling during the Second World War whale numbers did not rejuvenate.

Early international attempts at formally regulating the whaling industry were characterised by lack of both initiative and commitment, though there were obvious economic incentives for establishing a whaling regime. For example, the tremendous killing of Blue Whales in the early 1930s flooded the market with whale oil, quickly driving down prices. As David Francis writes, "it was evident to the whalers that national controls were not enough. Some kind of international regulation was necessary to stabilize the wild price fluctuations afflicting the industry."[27] The Convention for the Regulation of Whaling was signed in Geneva in 1931, and was later amended in 1937 as the *International Agreement for the Regulation of Whaling*. It did establish certain primitive guidelines: Right Whales and Bowheads, virtually extinct by this time, were exempted from further killing; prohibitions were placed against killing pregnant or lactating whales or calves; and paying whaleboat workers per whale taken was

---

[24] G. Kent, "Ocean Fisheries Management", in D. Orr and M. Soroos, ed., *The Global Predicament: Ecological Perspectives on World Order* (University of North Carolina, 1979), 232-248, 240.

[25] Webb, *op. cit.*, 83.

[26] See K. Watt, K. (1968), *Ecology and Resource Management* (New York: McGraw-Hill, 1968); C. Clark, "The Economics of Overexploitation", *Science*, 181:4100 (1973), 630-634; and the classic book by G. Small, *The Blue Whale* (New York: Columbia, 1971).

[27] D. Francis, *A History of World Whaling* (Toronto: Viking, 1990), 208.

outlawed. But the very nations currently expanding whaling operations (Japan, Germany, Chile, Argentina, and the U.S.S.R.) refused to sign. Negotiations throughout the 1930s, when modern whaling technology was employed at a frightening pace, resulted in little more than a "doubling of the kill [from 30,000 whales annually to 55,000] and an intensification of the hunt."[28] Regulation within the League of Nations framework amounted to what has been termed "regulated recklessness."[29]

It was not until 1946 that the *International Convention for the Regulation of Whaling,* which legally created the IWC, was signed (Japan would sign once it had resumed whaling in 1951). As this occurred in the immediate post-war international climate, it is not surprising that it was the result primarily of American initiative. The American government wanted to encourage global private investment in whaling and in the trade of whale products in as orderly fashion as possible. This would contribute towards bringing Japan into the larger economic system "while also exercising some control over the activities of the newest whaling nation, the U.S.S.R."[30] There is what Cherfas calls a "fine irony" in America's original insistence that the 1946 Convention not be a binding instrument: Article V allows states to escape from IWC decisions provided notice is given within 90 days of an objection. This is ironic because the U.S. would, only two decades later, begin to lead the anti-whaling coalition; but it is not surprising, since sovereignty was (and remains) the core formal principle upon which the United Nations system was and is based.[31]

The IWC is composed of one representative from each nation that is party to the ICRW. It has a small administrative Secretariat, a Technical Committee (which operates more like a forum for pre-voting consultation amongst appointed commissioners) which eventually recommends policy,

---

[28] D. Francis, *op. cit.*, 210.

[29] G. Jackson, *The British Whaling Trade,* (Hamden: Archon, 1978).

[30] R. M'Gonigle, "Economics, Ecology, and the Endangered Whale: Competing Foundations for Environmental Law and Politics", *Ecology Law Quarterly,* 9:1 (1980), 120-237, 132-133.

[31] For the complete text of the 1946 Convention see S. Lyster, *International Wildlife Law: An Analysis of International Treaties Concerned With the Conservation of Wildlife* (Cambridge: Grotius, 1985), 313-321. Utilising an interesting metaphor, Cherfas (*op. cit.*) writes that "America pulled the IWC's teeth in the first place, and ... now provides the IWC's only dentition", by way of the sanctions discussed below. To add to the irony, one nation demanding a more tightly binding original convention was Norway, whose whaling lobby would later cry foul when the IWC turned preservationist and which in 1992 unilaterally declared it would disregard the regime. In fact the unilateral Norwegian Whaling Act of 1929 was a precursor of international regulation (Francis, *op. cit.*, 208).

and a Scientific Committee composed of cetologists from member states. Members of the latter Committee, somewhat divorced from industrial interests, have played a limited yet important role in the evolution of the whaling regime, and this has been covered in great detail elsewhere.[32] Most significantly, the Committee was able in 1974 to persuade members to discontinue the devastatingly shortsighted "Blue Whale Unit" (BWU) system of quotas. The BWU system measured whales and allocated quotas in terms of expected oil yield relative to that produced by the Blue Whale, regardless of the state of particular populations. This was replaced by a more specific "maximum sustainable yield" system, which has since been further modified. But during the reign of the BWU in the 1960s, the whales diminished so rapidly that the level of kills could not even keep up to the absurdly high quotas set by the IWC.

In retrospect it would have been chimerical to expect anything but the continuation of large-scale whaling following the creation of the IWC, which was after all set up in a post-World War II context of scarcity "to provide for the proper conservation of whale stocks and the orderly development of the whaling industry."[33] (That these two purposes would prove contradictory was not acknowledged for some time.) It is thus rather unfair to criticise the industrial collusion between the IWC and the whaling consortium, since it was perceived as perfectly legitimate. In short, the IWC was administered by the whaling industry itself. Its voting structure demanded a three-quarter majority vote to change the Convention. So long as the majority of members were whaling nations and government delegates to the IWC were prepared to represent domestic whaling interests, the IWC would remain an instrument with which the whaling industry legitimized and, to some degree, regulated itself. Conducive governmental representation was assured not only in the case of the Soviet Union, where civil society had little say in the matter regardless, but also in polities where such interests were powerful economic actors. This has been the situation especially in the Japanese case, where the whaling industry, as a subsector of the immensely powerful fishing industry, has received inordinate promotion and protection from the cen-

---

[32] See Peterson, *op. cit.*, for a discussion of what one could call this "epistemic community." This may not be an accurate identification, since the broader "community" of cetologists and marine biologists is quite large, and includes researchers with quite different ideological, if not epistemological, orientations towards whales and whaling.

[33] The original text can be found in H. Degenhardt, "Organisations for the Preservation of the Living Resources of the Sea", in his *Treaties and Alliances of the World*, Fourth ed. (Detroit: Gale, 1986), 102.

tral government.[34] Japan's IWC Commissioner from 1951-1965 was K. Nakabe of *Taiyo Fishery Co. Ltd.*, arguably one of the most powerful Japanese companies even today, with direct links to the giant conglomerate *Mitsubishi*.

As in many other international organisations, voluntary membership has always instilled a sense of pragmatism in the proceedings of the IWC, though this was more evident before the adoption of pro-moratorium policies by the majority of member-states.[35] Provided they give sufficient warning, any member is entitled to opt out of a particular decision. Another structural feature of the IWC, however, would prove to facilitate its transformation: the Commission may be joined by any state which is willing to pay dues. By 1982, enough non-whaling nations had joined that the majority decision which imposed a complete ban on the taking of large cetaceans could be taken. This process of physical growth and normative transition began in earnest outside the IWC arena itself, at Stockholm, during the first United Nations Conference on the Human Environment in 1972.[36] Activists were able to use the image of the threatened whale as a "representation of the nonhuman living world" and this made an impression upon the attending army of diplomats and media. More specifically, a huge inflatable whale was paraded through Stockholm during the proceedings, a spectacle to be repeated at future IWC Annual Meetings.[37] What came out of that conference, personally delivered by its Secretary General, Maurice Strong, to the next annual meeting of the IWC, was Resolution 33, a call for a ten-year moratorium on commercial whaling, proposed by the United States and adopted almost unanimously, though later rejected by the whaling nations in the IWC as extreme.[38] Ten years (and thousands of whales) later, however,

---

[34] See Yutaka Hirasawa, "The Whaling Industry in Japan's Economy", in J. Schmidhauser and G. Totten, eds., *The Whaling Issue in US-Japan Relations* (Boulder: Westview, 1978), 82-114.

[35] In terms of binding agreements, the fear that extremism will spell the ruin of voluntary organizations is common and quite logical. J. L. McHugh, a former U.S. IWC Commissioner, said in 1978 that "[b]lind insistence on a total moratorium can be counterproductive by destroying a viable international mechanism of control achieved after long and hard bargaining." Quoted in R. McClung, *Hunted Mammals of the Sea* (New York: Morrow, 1978), 40. Iceland, by withdrawing from the IWC in 1992, is trying to prove his point.

[36] See W. Rowland, *The Plot to Save the World: The Life and Times of the Stockholm Conference on the Human Environment* (Toronto: Clarke, 1973).

[37] See L. Caldwell, *International Environmental Policy, op. cit*, 37 and 286; and D. Day, *The Whale War* (Vancouver: Douglas and McIntyre, 1987), 16.

[38] Rep. of the UN ConHE, UN Doc. A/Conf.48/14 Rev. 1. Japan voted against the Resolution, while the Soviet Union was not present in protest of the exclusion of East Germany from certain panels.

the moratorium would be accepted by the necessary three-quarter majority of IWC members. American involvement would be pivotal in the intervening period.

It is often said that every regime needs a leader. It comes as some surprise to many, however, that the United States took such a leading role in the development of a whale conservation regime. Indeed, it could be argued that the United Sates has been ideologically opposed to taking significant measures to reduce its contribution to global environmental problems. It has, however, played a leading role in wildlife conservation, especially regarding the establishment of CITES (February 1973 in Washington). The contradiction is not overly troublesome: American Administrations are characteristically reluctant to pursue foreign policy objectives which threaten to interfere with the activities of private enterprise, but have been quick to assert their international strength in areas which pose little threat to American industry. Proactive wildlife policy is largely the result of the concentrated energy of NGOs, to whom the whaling issue became so important in the early 1970s, though the strength of environmental lobbyists in the United States is much more evident in some issue-areas than others.[39] Despite a small whaling operation off of Richmond, California, closed in 1972, there was no significant national pro-whaling lobby in the United States and, as mentioned above, suitable replacements for whale products had been discovered. U.S. administrations, prodded by growing public pressure articulated by conservationist groups, scientists, and even labour unions, would lead the fight against whaling both by participating in the IWC arena and resorting to less diplomatic, or more precisely more economic, means of statecraft.

The latter involved the essential 1970 ban on imports of products derived from endangered Great Whales, and the enactment of relevant legislation which threatened the commercial interests of whaling nations.[40] Two amendments to older American legislation are of particular significance. The first relies on threatening access to what has always arguably been America's greatest asset, its internal market. The *Pelly Amendment to the Fisherman's Protective Act* of 1967 gives the President the option of banning any or all fisheries imports from a certified country; it was originally enacted "in response to Danish depletion of North Atlantic salmon in violation of the total ban put on that stock by the *International*

[39] See especially M'Gonigle, *op. cit.*, 190.

[40] See W. Schevill, ed., *The Whale Problem: A Status Report* (Cambridge: Harvard, 1974).

Convention for the Northwest Atlantic Fisheries."[41] The second is commonly known as the *Packwood/Magnuson Amendment to the Fishery Conservation Management Act* of 1976, which blocks access to U.S. coastal waters to nations that are violating bans on killing endangered marine mammals. This is of great significance since Japan takes some 75% of the bottom fish caught in Alaskan waters, worth around $750 million annually; and in 1983 the U.S. withheld 171,000 tonnes from Japan's fish allocations within the American 200-mile coastal zone in retaliation for Japan's initial objection to the 1982 moratorium.[42] In both cases, the Secretary of Commerce must certify offending nations, though the latter sanction is supposed to be mandatory while the former remains at presidential discretion.

The threat and, in rare cases, partial implementation of these sanctions have been credited with success in slowing down the pace of commercial whaling. Achievements at least partly resulting from the threat or implementation of sanctions include the changed policies of the Republic of Korea (banned the use of "cold harpoons" [i.e., harpoons without explosives, which damage less meat but result in considerably more suffering for the whale] in 1980); Taiwan (officially ended "pirate" [non-IWC regulated] whaling in 1980); China (banned whaling in 1982); Chile, Korea and Peru, which joined the IWC in 1979 (though it is often said these nations did so only for the right to legally sell whale meat to Japan, which in 1979 agreed to stop importing meat from non-IWC nations); Spain (observes fin whale quota, 1979); and Norway (ceased commercial whaling in 1987 after U.S. Commerce Secretary Baldridge certified Norway for continued catches on minke whales).[43] There is an irony at work here, given the American track record on environmental protection and industrial excess: the largest single polluter has the economic means to help save conserve wildlife at least partly because of the affluence generated by the very source of that pollution!

American governments clearly believed the whaling issue deserved priority. However, we would be remiss not to mention the economic threat posed by nonstate actors, such as the Animal Welfare Institute, who initiated consumer boycotts against whaling nations' products --

---

[41] R. Haskell, "Abandoning Recent Whale Conservation Initiatives in Japan Whaling Association v. American Cetacean Society", *Harvard Environmental Law Review*, 11:2 (1987), 551-91, 558-59.

[42] See Cherfas, *op. cit.*, 128; Haskell, *op. cit.*, 561; Lyster, *op. cit.*, 35.

[43] See D. Caron, "International Sanctions, Ocean Management, and the Law of the Sea: A Study of Denial of Access to Fisheries", *Ecological Law Quarterly*, 16:1 (1989), 311-354; and Day, *op. cit.*, 81-83.

in particular Japanese -- within the United States. This coincided with growing resentment of Japan's protectionist trade policies, and a perception that Japan was flooding the United States with foreign products. So the whaling issue became at some point subsumed by a larger, anti-Japanese movement and fear concerning America's industrial future. Indeed, in addition to lobbyists working in Washington, the Japan Whaling Association (JWA) at one point contracted a public relations firm in the U.S. to deal publicly with this controversial issue.[44]

There are notable exceptions to the record of American leadership. The ongoing debate over subsistence whaling periodically "neutralised" American determination, especially in 1977-78, until it was agreed that aboriginal whaling (in Alaska, Greenland, and Siberia) could take place provided Bowheads were not endangered and the hunters involved maintained IWC quotas. The political problem the U.S. faced was an acute one: could they actually stop the Alaskan Inupiat Eskimos from subsistence hunting if they accepted the IWC Scientific Committee's call for a total ban on Bowhead kills? Apparently, the decision was made that, though the Eskimos were expected to compromise previous takes, they would not be stopped altogether. The controversy over aboriginal subsistence whaling is a heated one beyond the scope of the present chapter.[45] But it alone does not explain the temporary change of American heart in the late 1970s: as Day reports, soon after his replacement in 1980

---

[44] *New York Times* June 6, 1974:7. See also R. Mandel, "Transnational Resource Conflict: The Politics of Whaling", *International Studies Quarterly*, 24:1 (1980), 99-127; and D. Hill, "Vanishing Giants", *Audubon*, 77 (1975), 56-90.

[45] During ths period the Americans were reduced to bargaining with the Japanese in order to obtain permission for the Alaskan hunt; as Day puts it, the administration was caught "between the Save the Whale lobby and the Save a Vote in Alaska lobby", and the general trade-off seemed to be one Bowhead for 200 Sperm whales (Day, *op, cit.*, 38). Aboriginal whaling, though still controversial, is regulated by the IWC and continues in Siberia, where Aleuts take Gray Whales; in Alaska, where Eskimos take Bowheads; in Greenland, where islanders take Minke and Fins; and St. Vincent and The Grenadines, where islanders take occasional Humpbacks. The Portuguese, still whaling in the Azores region, claim this is subsistence whaling - though Portugal is not a current member of the IWC. Interview with Ray Gambell, Secretary, IWC, April 22, 1992. There have been problems proving the validity of subsistence whaling; for example, the Siberian operations were at one point exposed as nothing more than operations to provide food for Soviet minke farms (Day, *op. cit.*, 73-77). The indigenous whaling issue is as legally as it is ethically complicated. For an argument supporting the right of indigenous whaling see N. Doubleday, "Aboriginal Subsistence Whaling: the Right of Inuit to Hunt Whales and Implications for International Environmental Law", *Denver Journal of International Law and Policy*, 17:2 (1989), 373-393; and for a reply see A, D'Amato and S. Chopra (1991), "Whales: Their Emerging Right to Life", *American Journal of International Law*, 85:1 (1991), 21-62.

the U.S. Commissioner during this time, Richard Frank, was in fact hired as a pro-whaling lobbyist for the Japanese Commissioner Kunio Yonezawa.[46] And there has been a certain reluctance to implement the full possible weight of sanctions for various reasons.[47] Nonetheless, we may conclude that willingness on the part of successive American Administrations to take a high profile approach with the whaling issue was instrumental in the developments which ensued.

But it would be superficial to attribute changes in the IWC solely to American initiative. In the late 1970s and 1980s many nations joined the IWC after lobbying from the Japanese on one side and anti-whaling coalitions, including persuasive NGOs, on the other. Due to the potentially universal membership structure of the Commission, nonwhaling nations are free to join and, regardless of their international stature, they have a full vote. Thus even the small Seychelles delegation, funded largely by the conservationist Threshold Foundation,[48] was able to play a significant role after joining in 1979, proposing bans on Sperm Whale hunting and the establishment of an Indian Ocean moratorium (which, incidentally, was renewed at the 1992 Glasgow meeting). Patricia Birnie feels the "strong environmental policies the Seychelles brought to the IWC emerged as a forceful counterweight to the traditional attitudes of the developing member states that were dominated by economic and sovereignty considerations."[49] The impact of the new members cannot be underestimated, and the fact that they are largely southern or "less-developed" nations, such as Uruguay, Senegal, Monaco, Kenya, Egypt, and Costa Rica, is of analytic interest. As Birnie reports, many southern nations entered the IWC for seemingly ulterior purposes, such as public relations (many of them joined during negotiations for UNCLOS in the early 1980s), and limit their participation to a bare minimum.[50] As Figure

---

[46] Day, *op. cit.*, 35.

[47] See D. Wilkinson, "The Use of Domestic Measures to Enforce International Whaling Agreements: A Critical Perspective", *Denver Journal of International Law and Policy*, 77:2 (1989), 271-292.

[48] See M'Gonigle, *op. cit.*, 190, n2. Seychelles is an island in the western Indian Ocean region; it gained independence from Britain in 1976, is governed by a fairly closed political system, and has 65,000 inhabitants. The Americans maintain a large satellite tracking station there.

[49] P. Birnie, "The Role of Developing Countries in Nudging the International Whaling Commission from Regulating Whaling to Encouraging Non-consumptive Uses of Whales", *Ecology Law Quarterly* 12 (1985), 937-975, 957.

[50] She elaborates: "Dominica never even appointed a Commissioner; Jamaica did so for only one year; Egypt and several other countries merely sent a delegate from their local embassy for a single day to cast a crucial vote; and many countries still have not paid their

5.2 indicates, many of these nations have quit the IWC since the establishment of the moratorium. This hardly matters to the whales whose lives were saved by their participation; but it does suggest we should avoid simply ascribing Southern behavior in this area solely to altruistic or idealistic motivation. Indeed, the claim is made that other nations are planning to join the IWC in order to vote against the moratorium in an effort to obtain Japanese development assistance.[51]

Other longstanding members contributed to the changed attitude of the IWC: Australia's delegation, pushed by domestic opposition, slowly turned to oppose commercial whaling, and Mexico, which declared Scammons Lagoon (a tourist attraction which benefits from an annual Gray Whale migration) a whale sanctuary in 1974, also added vocal impetus to conservationist efforts. So many members, both new and old, supporting within a legal framework a moratorium on the consumptive use of whales, presents a strong image of a 'normative snowball'. At present, there are some 36 members in the IWC, including China, Ecuador, France, Germany, Oman, the U.K., and many others. (Canada, which banned commercial whaling in 1972, dropped out of the IWC in 1982, disappointing many whale enthusiasts, though it maintains observer status and contributes to the work of the Scientific Committee.)[52]

The IWC was only one, if the principal one, of a number of international organisations which played a role in its own normative transition. The General Assembly of the IUCN declared in 1981 that "the view is now held by the people of the majority of nations of the world that for scientific, ecological and aesthetic reasons whales should no longer be killed for industry."[53] We should note that the IWC is in fact strengthened by a network of other international institutions, principally by the environmentalist agendas of several United Nations bodies; the United

---

contributions." *Ibid.*, 962. As we shall see later, Japan has been accused of actually bribing developing states to join the IWC in order to vote **against** the moratorium.

[51] Kate O'Connell, international conservation director of the British Whale and Dolphin Conservation Society, claims that "[t]he quickest way to get guaranteed aid from Japan is to say, 'I'm interested in the whaling issue.'" Quoted in A. Pollack, "Commission to Save Whales Endangered, Too", *New York Times*, May 18, 1993, C4.

[52] Thanks to Dan Goodman, Oceans and Fisheries Canada and Canadian representative to the IWC, for discussions in April 1992 and January 1994.

[53] 15th Session, Christchurch, New Zealand, Oct. 11-23, 1981; quoted in B. Raster and B. Simma, ed., *International Protection of the Environment: Treaties and Related Documents*, Second Series (New York: Oceana, 1989), 20.

Figure 5.2: Membership in the IWC, 1971-1992

Year	IWC Quota	Members
1972-73	42,500	14: USA, Britain, Mexico, France, Argentina, Australia, Canada, Panama, Denmark, Japan, USSR, Norway, Iceland, South Africa
1973-74	40,979	
1974-75	39,864	15: Brazil joins
1975-76	32,578	
1976-77	28,050	16: New Zealand joins
1977-78	23,520	17: Netherlands joins
1978-79	20,428	
1979-80	15,653	23: Seychelles, Sweden, Chile, Peru, Spain, Korea join
1980-81	14,523	24: Switzerland and Oman join; Panama withdraws
1981-82	14,233	32: Jamaica, St. Lucia, Dominica, Costa Rica, Uruguay, China, St.Vincent, India, Philippines join; Canada withdraws
1982-83	12,263	37: Egypt, Kenya, Senegal, Belize, Antigua, Monaco, Germany join; Jamaica, Dominica withdraw
1983-84	9,393	39: Finland, Mauritius join
1984-85	6,623	
1985-86	0[a]	41: Ireland, Solomon Islands join
1986-92	0	36: Ecuador, Venezuela join; Solomon Islands, Mauritius, Belize, Egypt, Philippines, Uruguay, Dominica, Jamaica, Iceland withdraws

a: discounting aboriginal and scientific whaling quotas

Source: D. Day, *The Whale War* (Vancouver: Douglas and McIntyre, 1987), 97; and documentation from the IWC main office in the U.K.

Nations Conference on the Law of the Sea[54] (UNCLOS); CITES (all cetaceans whose catch is regulated by the IWC are in Appendix One of the Convention); the EEC, which in 1981 banned the import of whale products for commercial purposes; and the important 1982 Convention on the Conservation of Antarctic Marine Living Resources (CCAMLR) which regulates the exploitation of relevant mammals and krill.[55] Agenda 21, the lengthy framework that emerged from UNCED, makes explicit mention of the importance of whale conservation and the role of the IWC and its Scientific Committee (Chapter 17:61, a and b).

As importantly, the participation of non-governmental organisations in the whaling case is by now famous. These groups included Friends of the Earth, Greenpeace, the International Society for the Protection of Animals, Project Jonah, the World Wildlife Fund, and many others, mainly of North American origin. The whaling issue became a rallying point for these groups, and figured prominently in many of their itineraries. They were able to capture public imagination by way of peaceful protest and the distribution of information. Much media attention was focused upon incidents that involved violence, such as the ramming of the notorious pirate ship *Sierra* by the renegade "radical environmentalist" Paul Watson, and the destruction of Icelandic whaling property in October 1986. Other bombings in Portugal and Spain followed. The topic of "monkeywrenching" is beyond the scope of this chapter, but we can with certainty conclude that the quest for cetacean survival has provided a strong stimulant.[56] We may also note the instrumental role played by

---

[54] Article 65 states that "States must abide by the regulations of the IWC, whether or not they are Parties to the Whaling Convention, except where they adopt stricter measures for the conservation of whales." This is particularly significant for states which do not belong to the IWC but ratify LOSC, and have allowed their national flag to be flown by "pirate" whalers in order to escape IWC restrictions. Lyster, *op. cit.*, 39; see also P. Birnie, "The Development of the International Regulation of Whaling: Its Relation to the Emerging Law of Conservation of Marine Mammals", Ph.D. Thesis, University of Edinburgh, 1979. Of course there are also maritime international organizations which serve functionalist roles above or beyond the UNCLOSC, such as the Intergovernmental Maritime Organization and the International Oceanographic Commission.

[55] On this see J. Barnes, "Legal Aspects of Environmental Protection in Antartica", in C. Joyner and S. Chopra, ed., *The Antarctic Legal Regime* (Dordrecht: Martinus Nijhoff, 1988), 241-268; and same, "The Emerging Convention on the Conservation of Antarctic Marine Living Resources: An Attempt to Meet the New Realities of Resource Exploitation in the Southern Ocean", in J. Charney, ed., *The New Nationalism and the Use of Common Spaces: Issues in Marine Pollution and the Exploitation of Antarctica* (New Jersey: Allanheld, 1982), 239-286.

[56] Day, *op. cit.*, 53-59; 132-13. Watson has apparently written a book: P. Watson and W. Rogers, *Sea Shepherd: My Fight for Whales and Seals* (New York: W.W. Noerton, 1982). I have not seen this book. Watson left Greenpeace in the late 1970s complaining of its bureaucratic

Greenpeace and other non-governmental actors in exposing both illegal and pirate whaling.[57] The extensive Japanese whaling empire worked in and out of the IWC's jurisdiction: Japanese companies have been accused of establishing, through the export of capital and expertise, pirate "whaling colonies", primarily in South America (Peru, Brazil, Chile) but also in Taiwan, South Korea, and the Philippines. In the absence of any international state-sanctioned system of surveillance which could efficiently cover such widespread operations, it was essential that non-governmental organisations, and their motivated members, were able to -- despite continuous confrontation with state authorities -- participate in the emerging whaling regime.

It is difficult to arbitrarily decide how much of an impact the NGOs had on state policy: for one of the first times, they demonstrated considerable lobbying skills in persuading non-whaling nations to join, their publicity stunts generated media coverage unprecedented in environmental politics, and they attracted large numbers of highly educated individuals which legitimated their cause in the eyes of many previously biased observers. While it needs no glamorization here, it is in retrospect a truly heroic tale, one which raises images of both assertive empowerment and professional sacrifice. Perhaps the best testament to the importance of NGOs in conservationist politics, both local and global, is their impressive growth and resilience. NGOs continue to play a large role at IWC annual meetings: though they are not given voting status they can

---

inertia. See C. Manes, *Green Rage: Radical Environmentalism and the Unmaking of Civilization* (Boston: Little, Brown and Company, 1990), esp. 107-111; and M. Bookchin, *Defending the Earth* (Montreal: Black Rose, 1991). Watson, maintaining his celebrity, has recently been involved in overfishing protests off the Canadian Atlantic coast; the *Sierra* is reportedly rusting somewhere, causing environmental damage in the process.

[57] The conventional distinction between illegal and pirate whaling is interesting: the latter is usually construed as all whaling which takes place outside the regulatory scope of the IWC; the former refers only to member-state violations of those regulations. This raises interesting jurisprudential problems for international law specialists, as well as the question -- tantalizing to regime theorists -- of what, exactly, a prohibitionary regime (which we discussed in the Second Chapter) really is. See P. Birnie, *Legal Measures for the Prevention of 'Pirate' Whaling* (Switzerland: IUCN, 1981); and C. van Note, *Outlaw Whalers: An Expose of Unregulated Whaling Around the World* (Whale Protection Fund, 1979). For American governmental investigation, see *Outlaw Whaling: Hearing Before the Committee on Commerce, Science, and Transportation,* United States Senate, 96th Congress (First Session on Whaling Operations Conducted Outside the Control of the IWC), June 22 (Washington, 1979); *Review of the 34th IWC Meeting,* Hearing Before the Subcommittee on Human Rights and International Organizations of the Committee on Foreign Affairs, House of Representatives, 97th Congress, September 16 (Washington, 1982); and, for a portrait of a notorious outlaw whaler, see P. Evans, *Ari: The Life and Times of Aristotle Socrates Onassis,* (London: Jonathan Cape, 1986).

sit in as observers and engage in considerable "arm-twisting" in this capacity.[58] The whale and interested NGO's have, in fact, established a unique symbiotic relationship, since the issue helped galvanize and popularize the peripheral promotional efforts of the latter. For example, as two activists explain, "Greenpeace cut its teeth on whaling."[59] Such NGOs, to employ Young's terminology, are organisations which became instrumental components of the evolving social institution surrounding the whaling issue.

4. THE LEVIATHAN: SYMBOLIC POWER AND IMPLICATION

Another actor, in fact the one most closely involved in these developments and yet unable to take direct political action, deserves at least some analytic attention. It may be suggested that the unique character of whales themselves helped get the 'normative snowball' mentioned above rolling. There are three factors, in particular, which make whales different. Firstly, and most obviously, the sheer size of the Great Whales, and especially the Blue Whale, the largest creature to have lived on earth, is literally hard to overlook. It seems, in terms of wildlife conservation, that size and public curiosity (and empathy) are positively correlated. Elephants, rhinoceroses, gorillas, polar bears, and other large mammals, which have been quaintly labelled "charismatic megavertebrates", capture our imagination perhaps because they make us, unaccustomed to humility

58 Interview with Ray Gambell, Secretary of the IWC, April 22, 1992. A list published by the Commission of attenders of the 41st Annual Meeting (June 1989 in San Diego) included many diverse organizations: the African Wildlife Foundation, Assembly of Rabbis (disguised "eco-detectives" according to Day [*op. cit.*, 21-22]), Fundacao Brasileira para a Conservacao da Natureza, Save the Children, and the Women's International League for Peace and Freedom.

59 C. Plowden and S. Shoup (1990), "Whaling", in *Greenpeace* Jan/Feb 1990, 10. See also R. Hunter, *Warriors of the Rainbow: A Chronicle of the Greenpeace Movement* (New York: Holt, Rinehart and Winston, 1979); and Hunter and R. Weyler, *To Save a Whale: The Voyages of Greenpeace* (Vancouver: Douglas and McIntyre, 1978). This issue-dependency has been criticized, for example by David Buzzelli of Dow Chemical: "...each environmental issue group has mastered the technique of attracting and locking-in media and public attention for its single cause. As a result, the public hears a loud, mixed, emotional overload. No wonder eyes glaze over during the nightly news broadcasts." "Key Environmental Issues for the 1990s and Beyond in Canada and the United States", *Canada-United States Law Journal*, 18 (1992), 11-21, 11.

as we are, seem small.[60] Of course, the whale's size has historically worked against it, since it represented to enterprising whalers a virtual bounty of natural resources. That mammals larger than dinosaurs which have roamed the seas for millions of years were threatened with extinction by something as ephemeral as the sunk costs involved in operating a whaling fleet says much about human economics and consumption; but the fact that the Leviathan has attracted our imagination for so long says something else.

Secondly, there is the intriguing question of whale intelligence. Anthropocentric verdicts, based on comparisons with human intelligence, vary, usually according to one's perspective on whaling: the Japan Whaling Association is prepared to equate whales with domesticated cows; cetologists quibble, but there is consensus that whales have extraordinary communication skills; and extremists argue whales, who have lived in relative harmony with their environment for ages, may even be wiser than *Homo sapiens*.[61] Those that argue the latter group is romanticising miss the point; many humans have formed a mental bond with the whale, as symbol, precisely because it is perceived as highly intelligent. Such bonding, which has occurred within the non-conducive social context of western scientific and industrial civilization, is significant in terms of human-nature interaction, and it hints at the intellectual roots of the ecological sanity of many non-industrial communities. It is indeed ironic that attempts to somehow measure whale intelligence in the tradition of positivist science have lead to a greater normative understanding of the effects of our own.

Most importantly from the perspective of international relations, whales are migratory: they escape the jealous confines of sovereignty which have been imposed upon earth with the advent of human organisation. As we have seen, environmental problems in general challenge the logic of the nation-state system. This is especially so in regard to questions of the Commons, where the risk-oriented decision-making process cannot be a unilateral affair. The political economy of whaling is a distinct example of the complex character of such a phenomenon, and is so because whale nations are, for the most part, transient collectives. This does not mean, of course, that the sovereignty question loses significance;

---

[60] T. Foose, "Riders of the Last Ark: The Role of Captive Breeding in Conservation Strategies", in L. Kaufman and K. Mallory, ed., *op. cit.*, 141-165.

[61] See K. Fichtelius, and S. Sjolander, *Man's Place: Intelligence in Whales, Dolphins, and Humans* (London: Gollancz, 1973). Most texts on cetaceans devote some space to the question of intelligence.

on the contrary, calls for the sanctity of cultural sovereignty have characterized the Japanese whaling industries' rhetoric. But the "Common" character of whales remains highly symbolic. In a famous statement made in 1946, U.S. Secretary of State Dean Acheson waxed poetic about whales: "[t]he world's whale stocks are a truly international resource in that they belong to no one single nation, nor to a group of nations, but rather they are wards of the entire world."[62]

5. The Future of Whale Conservation

The whales threatened by hunting today are very small compared to their formerly harassed but now protected relatives. These are the more populous small cetaceans, the Minke, Bottlenose, Beaked, Pilot and Killer whales. Iceland's withdrawal from the IWC in 1992, and Norway's unilateral (but entirely legal, given the IWC's framework) decision to resume commercial whaling of Minke whales (*balaenoptera acutorostrata*) in 1993 has revived the international debate over both the ethics of whaling and the sovereignty of whaling nations.[63] The latter problem is illustrated by the apparent international impotence to change the traditional "drive fishery" technique of mass Pilot Whale slaughter in the Faeroe Islands.

It is, as Lyster writes, unclear whether the 1946 Convention covers small cetaceans.[64] At present, there are obvious lapses in the observational powers of an organization like the IWC. Estimating whale stocks remains a haphazard endeavour. Monitoring catches is equally arduous, since the IWC cannot even appoint its own inspectors but must rely on those supplied by various members, who then board declared whaling vessels.[65] In other words, the nations are to watch each other; this leaves

---

[62] Quoted in Francis, *op. cit.*, 207. How creatures that have survived quite well on their own for over thirty million years became "wards" of humanity is not yet scientifically understood!

[63] At the last IWC General Meeting, Norway's representative stated that Norway "no longer accepts what she perceives as cultural imperialism imposed by the majority of the members of the IWC on the local communities of the nations and peoples who want to exercise their sovereign cultural right to be different." IWC, *Chairman's Report of the 44th Annual Meeting of the International Whaling Commission* (Cambridge: IWC, 1992), 11.

[64] *Op. cit.*, 22.

[65] These arrangements are often strengthened by bilateral agreements, such as the *Exchange of Notes Constituting an Arrangement Relating to an International Observer Scheme For the Whaling Station at Cheynes Beach in Western Australia*, UN Doc. 17495, UN Treaty Series Vol. 1124 (1979), 43-48; or the *Agreement Between the USA and Japan Concerning an International Observer Scheme*

a lot to faith (and makes the functional vigilance of watchful groups like Greenpeace significant). Some members are calling for an extension of the IWC's explicated functions to include monitoring and regulating populations of small whales, even if it must be done on a regional (and, thus, more expensive) basis. The Americans may be pushing for this extension of IWC jurisdiction in the near future.[66] Martin Harvey, the current Executive Officer of the IWC, is quick to point out the considerable organizational and financial difficulties decentralization would entail, especially if current membership levels do not rise.[67] It also raises questions about sovereign jurisdiction. There is an example of a regional whaling commission, to which Peru, Chile and Ecuador belonged, but it was in fact very similar to the original IWC (in other words, it facilitated whaling) and disintegrated when these nations joined the IWC in the late 1970s. This was the Permanent (!) Commission of the Exploitation and Conservation of the Marine Resources of the South Pacific. Meanwhile, renegade Iceland has established a North Atlantic Marine Mammal Commission, though it is not dealing with large cetaceans. Aside from the logistical challenges, an expansion in the IWC's mandate to include small cetaceans raises interesting ethical questions: if it is wrong, for reasons other than ensuring their future consumption, to kill Great Whales, what does this imply about small whales, marine mammals, other forms of wildlife? Where is the line drawn? We can expect a flurry of further debate as this issue surfaces in the future. But other immediate threats exist.

In light of the learning experience of whaling in this century, it is both tragic and absurd that it continues under the guise of "scientific whaling." Clearly, we learn more about whale behavior by studying living mammals, and it is the market for whale meat -- limited mainly to Japan -- that fuels continued whaling. Though the numbers taken are small compared to the heady days of commercial whaling, it is true that they exceed what any cetologist might demand for research purposes, and attempts to introduce stricter guidelines "tying such permits to demonstrably productive scientific programmes not amenable to other research methods" have been abandoned for milder standards.[68]

For Whaling Operations From Land Stations in the North Pacific Ocean, UN Doc. 14640, UN Treaty Series Vol. 1148 (1979), 444-405.

66 Interview with John Knaus, IWC Commissioner for the United States, April 22, 1992.

67 Interviews, April 23, 1992 and September 22, 1992.

68 This proposal was put to the IWC in 1985 by Sweden and Switzerland; see M. Forster, "IWC: Trying Hard?" *Environmental Policy and Law*, 15 (1985), 8-9. Despite scientific interest, the Japanese market for whale meat remains a pivotal factor: writes Cherfas, with respect to Iceland's proposed scientific whaling catch: "there are not enough Icelanders to eat all the

Other threats persist. One is the continuation of driftnet fishing, one of the least sustainable methods of fishery harvesting ever devised, which inadvertently kills dolphins, porpoises, and other small whales. In 1989 a United Nations General Assembly resolution called for a global moratorium on driftnet fishing by mid-1992. Japan has agreed to halt this practise by the end of 1992, and a Convention to the Prohibition of Fishing with Long Driftnets in the South Pacific has been created; Taiwan, South Korea and Mexico continue to be problem states in this respect.[69]

Yet another threat related to human extractive activity is the diminution of food supply by the increased harvesting of krill from the Southern Ocean of Antarctica. The krill question demonstrates just how interdependent ocean ecology really is: whaling reduced the larger krill consumers, which in turn led to increased populations of "seals, seabirds, squid, and those baleen whale populations that had not been exploited."[70] Conversely, "any success in efforts to ... [encourage] the growth of stocks of great whales will have important implications for arrangements governing the harvest of renewable resources, such as krill, in the Southern Ocean."[71]

However, and provided an unlikely major reversal of the moratorium does not take place, the ultimate threat to the whale comes not from the harpoon, but from the excesses of industrial societies which -- beyond habitual overfishing -- pollute aquatic environments not only in chemical, but aesthetic and, even, acoustic, terms. As mentioned above, the greatest threat to wildlife has always been related to the destructive alteration of its habitat, and the oceans prove particularly troublesome in this context precisely because they come under the jurisdiction of no single sovereign power. They are, indeed, the commons, though we have defined this in an anthropocentric manner to mean our commons, susceptible to the vicissitudes of our exploitation. The negotiations at UNCLOS III made this clear: the conservation of marine biota was

whales they plan to kill for science" *op. cit.*, 183. See also *Greenpeace* 13 (July 1988), 19-20. Iceland did not conduct a scientific hunt after 1989, but Japan and Norway continue to take Minkes in this capacity.

[69] Toronto *Globe and Mail*, Nov. 27, 1992: 16; see D.M. Johnston, "The Driftnetting Problem in the Pacific Ocean: Legal Considerations and Diplomatic Options", *Ocean Development and International Law*, 21:1 (1990), 5-40.

[70] G. Knox, "The Key Role of Krill in the Ecosystem of the Southern Ocean with Special Reference to the Convention on the Conservation of Antarctic Marine Living Resources", *Ocean Management* 9, (1984), 113-156, 143.

[71] O. Young, *International Cooperation: Building Regimes for Natural Resources and the Environment* (Ithaca: Cornell, 1989), 100.

"considered an economic measure -- to assure continued availability of the resource -- rather than an ecological measure -- to maintain the health of the marine environment."[72] Ultimately, the extensive normative transition necessary to sustain the hydrosphere, and with it life on this planet, may be unattainable.

Protecting whales from harpoons will be rather senseless if their environments do not receive commensurate protection. This, of course, is a daunting task, but there may be a window of symbolic opportunity here. With regard to linking ocean habitat with the plight of the whale, the establishment of the Indian Ocean Whale Sanctuary, which received a ten-year lease on life at the IWC meeting in Glasgow in 1992, is a glimmer of hope. This sanctuary may even become continuous with the Antarctic sanctuary proposed by France this year (the Scientific Committee had yet to consider the proposal). Though other national sanctuaries exist, these multilateral agreements could be of great long-term importance. Symbolically, and with some stretch of imagination, this is ultimately about creating geopolitical space for whales. It also contributes to the establishment of *opinio juris* (the psychological component of international customary law), and by extension an "emergent entitlement of whales -- not just 'on behalf of' whales -- to a life of their own."[73] Preserving their environment (in our own best interest anyway) becomes obligatory.

6. Conclusion (and Postscript)

Fears that Iceland's withdrawal signals the beginning of the end of the current whaling regime are unwarranted at this time: world public opinion and, more concretely, the continued threat of American sanctions, make them so. However, we should not expect the IWC to expand as an organisation in functional terms, to include the conservation of the smaller whales discussed above. It may even be the case that an alternative organisation, explicitly preservationist, will be developed, though this would be extremely difficult to achieve with any sort of inter-state consensus given the non-universal nature of anti-whaling sentiment. However, it may be argued that there is sufficient normative consensus to maintain the present non-consumptive philosophic outlook in the IWC

[72] M. Glassner, *Neptune's Domain: A Political Geography of the Sea* (Boston: Unwin Hyman, 1990), 72.

[73] D'Amato and Chopra, *op. cit.*, 23.

and the broader whaling regime. In terms of international antagonism, the lines are clearly drawn. One of the American delegates to the 1992 IWC Conference was Michael Sutton, a senior program officer at the World Wildlife Foundation; one of the Japanese delegates was Kazuo Shima, the deputy director general of the Japanese Fisheries Agency.[74]

There is a marked tendency to identify international organisations in particular as the nexus point between systemic change and new avenues of thought. This may be the stuff of overstatement. But we can certainly argue that such organisations do offer opportunities for actors to communicate and reach agreement while retaining their sovereign priorities. They may also provide a medium -- a billboard -- for the limited dissemination of viewpoints and demands emanating from both epistemic communities and civil society. In short, they may be necessary but insufficient.

The IWC should be seen in this less dramatic light. No doubt, the whaling regime represents the end result of years of bargaining within the IWC. But to avoid the politics exogenous to the international organisation itself would be to miss the point, if not the story; and the biological and symbolic role of the Great Whales themselves have played an inspiring, if costly to them, role. Similarly, the future debate will be about economics, sovereignty, and environmental philosophy.

As regards the latter, change is a relative thing. At a "Workshop on Humane Killing Techniques for Whales" held at the IWC Meeting of 1980, the group noted that the "development of a means of achieving a rapid, painless death would also and incidentally increase the efficiency of the whaling operation and improve the quality of the meat obtained (by reducing the stress caused to the whale)." Twelve years later, the "Humane Killing Working Group" at the IWC Meeting in Glasgow recommended that appropriate governments and national agencies "develop procedures for ... representative samples of blood, brain and other tissues from selected animals, to allow assay of stress indicators and other physiological parameters in animals killed in whaling operations."[75]

---

[74] S. Weisman, "Whale: Food For Deep Thought, or Just Food?" *New York Times*, April 4, 1992, A6.

[75] IWC, 1992 Report *op. cit.*, 6-7.

POSTSCRIPT[76]

Despite an elaborate publicity campaign waged by the Japanese fishing industry (which included giving free samples of whale meat to delegates), the International Whaling Commission (IWC) has once again voted to upkeep the global moratorium on large-scale commercial whaling adopted in 1982 and in effect since 1988. This year's annual meeting was held mid-May in Kyoto, Japan, where the hosts argued that enough of the relatively small minke whales exist to justify some level of commercial whaling. In the IWC, to which some 40 countries now belong, a two-thirds majority is needed to overturn the moratorium; and the whaling resumptionists did not muster the necessary votes.

The results are not, however, a total success for anti-whaling enthusiasts. A proposal to declare the Antarctic a whale sanctuary has been all but scuttled (it will be discussed next year in Mexico), Norway remains determined to rejuvenate its relatively small whaling industry, and so-called "scientific" whaling will continue. Iceland, which pulled out of the IWC last year, is determined to pursue an autonomous policy on whaling, and no one seems sure what the cash-strapped Russians, whose whaling industry once rivalled that of the Japanese in the old days of excessive catches, have in mind. But the moratorium is notable for its longevity and is highly symbolic for many environmentalists.

In this context, it is interesting to note the comments of the irate Japanese IWC Commissioner, Kazuo Shima, following this year's Meeting: "It is really deplorable that the IWC has become an organization to protect the rights of whales."[77] One thing is certain: a great deal of controversy will continue to surround this international issue in the future, and the central and largely ethical debate between *preservationists* and *conservationists* continues.

---

[76] For a lengthier version of this postscript, see "The IWC: Holding Firm on Whaling", in *Sea Wind: Bulletin of Ocean Voice International*, 7:1 (1993), 22-23. This case study of whaling will be considerably expanded in a forthcoming book, *The International Politics of Cetacean Survival* to be completed soon by the present author.

[77] *The Globe and Mail*, May 15, 1993.

CHAPTER SIX

RIVERS: DEVELOPMENT, GEOPOLITICS, AND THE ENVIRONMENTAL ALTERATION OF THE MEKONG[1]

"Global concern about the social and environmental costs of development is increasing and this is tending to hamstring governments in the developing world and prevent them from taking action to improve social conditions."
- editorial, *Mekong News: Newsletter of the Interim Mekong Committee*, 1991[2]

1. INTRODUCTION

Conflict over "resource domains"[3] has been a recurring theme in the history of international relations, while environmental concerns are increasingly popular in contemporary world politics. Quite understandably, then, *international resource domain management* has assumed newfound importance in the study of global and regional international

---

1 An earlier version of this chapter, entitled "Regional International Organization: Southeast Asia's Mekong River Politics" was presented to the "Asia in the 1990s: Meeting and Making A New World" Conference, Queen's University, Kingston, October 1993.

2 Bangkok: Public Information Unit of the Mekong Secretariat, 10:2 (May, 1991), page 2.

3 Defined by Marvin Soroos as "an area or region ... that encompasses something that is of use to human actors and over which some form of jurisdiction can potentially be exercised." "Conflict in the Use and Management of International Commons", in Jyrki Kakonen, ed., *Perspectives on Environmental Conflict and International Relations* (London: Pinter, 1992), 31-43, 42n1.

organization. It is perceived as a double benefit to troubled regions such as Southeast Asia, providing an incentive for international political cooperation and the infrastructure necessary for rapid industrialization. It also creates controversy as those opposed to the social and environmental effects of this management adopt strategies of resistance and publicize possible negative factors. These contrary perspectives will emphasize different consequences of what we may label, in an effort to avoid the heavy connotations associated with the word *development*, "environmental alteration."

Of course, environmental alteration is to some extent inevitable, since human-nature interaction changes both. What is of concern is in essence the severity of such change. The development of a hydroelectric dam, for example, often converts flowing rivers into lacustrine ecosystems by eliminating natural water currents. The effects of such alteration, on local inhabitants, wildlife, vegetation, and geophysical structures, are often overlooked since development at any cost has been the dominant paradigm of industrializing societies. This is changing as institutions such as the World Bank (IBRD) are slowly prodded by both southern resistance to such projects and northern environmental activism into adopting policies sensitive to the need for natural conservation. But this is hardly countering what is an overall trend toward increased resource exploitation and attendant human displacement in development planning. This is especially so in Southeast and Pacific Asia, since those nation-states are so heavily dependent on exporting natural resources. It is also especially evident in nation-states with rapidly increasing electrical energy demands, such as Thailand.

There are many contributing factors which drive this socio-economic process: overpopulation, elite power structures, foreign investment, structural inequity and debt cycles within the international system, and a technocratic bias towards the development megaproject. At the same time we will also see a concerted effort, at the local and international level, to minimize the potential damage caused by development projects to both the environment and, in the case of shared resources, to official relations. This is not always a successful venture, as demonstrated by the current conflict between Hungary and Slovakia over the tortuous, and probably disastrous, construction of the Danube River dam system.[4]

---

[4] See in particular J. Galambos, "Political Aspects of an Environmental Conflict: The Case of the Gabcikovo-Nagymaros Dam System", in J. Kakoneen, ed., *op. cit.*, 72-95. Hungary and Slovakia are currently awaiting the outcome of a World Court consideration of the question, which has serious implications for Slovak-Hungarian relations. The Slovaks have diverted the

However, cooperative arrangements are at least as likely, and there are several prominent African examples, including the Unduga Group (Nile Basin), the Lake Chad Basin Commission, and the Niger Basin Authority.[5] The recent Israeli-PLO peace agreement showcased in Washington raises hopes for a permanent water sharing agreement in the Middle East.[6] The need for widespread effective intergovernmental management is clear: there are more than 200 river basins shared by two or more countries. One of them, the Mekong, offers an interesting and important example of a political space where international and substate relations, development, and environmental concerns converge.

2. INTERNATIONAL RELATIONS AND RIVERS

Cooperation over river resources may be ephemeral, while the human need for fresh water resources -- lakes, rivers, tributaries, and aquifers -- is eternal. The primary need for water, and its many provisions, may bring out the worst in us yet. Jon Martin Trolldalen writes of three principal types of international environmental conflict, which he calls the ABC model: A) incompatible goals related to access to the river; B) externalities created utilizing the resource, such as the salination that results upstream from dam operations; and C) the result of externalities from other activities affecting the river system (such as eutrophication resulting from chemical discharge by factories).[7] As Peter Rogers writes, "water problems usually involve portraying a neighbouring nation or people as an antagonist, which tends to intensify popular emotions, cluster water issues with other historical grievances, and favor the combative

river in order to operate a dam designed for joint management on their own; as a result, "Hungary lost control last October over the flow into one of its most important wildlife habitats and sources of drinking water." *The New York Times*, Sunday July 11, 1993, page 10.

5 See F. Visser, "Recent Developments in the Joint Management of International Non-Maritime Water Resources in Africa", *The Comparative and International Law Journal of Southern Africa*, XXII:1 (1989), 59-92.

6 See A. Cowell, "Hurdle to Peace: Parting the Mideast's Waters", *The New York Times*, Sunday., Oct. 10, 1993, page 1. For a pre-agreement analysis see M. Lowi, "Bridging the Divide: Transboundary Resource Disputes and the Case of West Bank Water", *International Security*, 18:1 (1993), 113-138.

7 J. M. Trolldalen, *International Environmental Conflict Resolution: The Role of the United Nations* (Washington: World Federation for Environment and Development, 1992).

set of attitudes associated with zero-sum situations."[8] A more 'liberal institutionalist' or pluralist argument would assert that the practical experience of cooperation over resource domain management will foster enhanced regional security; some analysts stress the cooperative potential afforded by water needs, since neighbouring states are in essence obliged to invest in mutually vulnerable infrastructure.[9]

Neofunctionalists (see Chapter Two) might argue further that the institutions set up to manage common resource domains could prove to be the "spillover" mechanisms which induce greater and more complex cooperative arrangements in other areas. This view -- that of regional international organizations as catalysts of regional integration -- may be before its time in the case of areas like South Asia or Southeast Asia, where acrimony remains a poignant feature of interstate interaction. But the softer argument, that resource domains should if possible be used as peace and confidence building mechanisms, remains. The post Cold War era is a decidedly disorderly one, and it will be characterized by the fusion of security and environmental concerns. From a simple international security perspective, any attempt to cool down the world's "hot spots" of potential or realized conflict such as Eastern Europe, the Middle East, Southern Asia, and the geographic focus of the present chapter, Southeast Asia, should be pursued with vigour. However, this perspective obscures sub-state actors, such as those directly affected by large scale development projects, and the environment itself. Insisting on our wide definition of environmental security, or what we have termed GES, puts people back in the equation.

---

[8] P. Rogers, "Economic and Institutional Issues: International River Basins", in Guy Le Moigre, et.al., eds., "Country Experiences with Water Resources Management: Economic, Institutional, Technological and Environmental Issues", World Bank Technical Paper No. 175 (Washington, 1992), 63-70, 63. See also P. Gleick, "Water and Conflict: Fresh Water Resources and International Security", *International Security*, 18:1 (1993), 79-112; M. Falkenmark, "Fresh Waters as a Factor in Strategic Policy and Action", in A. Westing, ed., *Global Resources and International Conflict: Environmental Factors in Strategic Policy and Action* (Oxford: Oxford, 1986), 85-113; and J. Starr, "Water Wars", *Foreign Policy*, 82 (Spring 1991), 17-36.

[9] D. Deudney, "The Mirage of Eco-War: The Weak Relationship Among Global Environmental Change, National Security and Interstate Violence", in I. Rowlands and M. Greene, *Global Environmental Change and International Relations* (London: Macmillan, 1992), 169-191; 183.

Table 6.1: International Rivers and Intergovernmental Conflict

River System	Countries Involved	Main Cause[a] of Dispute
Nile	Egypt, Iraq, Sudan	water flow
Euphrates, Tigris	Iraq, Syria, Turkey	dams, water flow
Jordan, Litany	Israel, Lebanon	water flow
Yarmouk	Jordan, Syria	water flow
Indus, Sutlei	India, Pakistan	irrigation
Ganges	Bangladesh, India	siltation, flooding
Mekong	See this study	water flow, dams
Parana	Argentina, Brazil	dam, flooding
Lauca	Bolivia, Chile	dam, salination
Rio Grande	Mexico, USA	salination, pollution
Great Lakes	USA, Canada	pollution, diversion
Rhine	four Rhine states	pollution
Elbe	Czech, Slovakia, Germany	pollution
Szamos	Hungary, Romania	pollution
Danube	Czech, Slovakia, Germany, Hungary	dam, water flow

a: though the "main" causes are listed here, it is never this simple, if only because so many causes of conflict are interrelated.

Source: J. M. Trolldalen, *International Environmental Conflict Resolution: The Role of the United Nations* (Washington: World Federation for Environment and Development, 1992), 62.

International law has been applied to the question of shared river resources, but we should hesitate before placing much emphasis on its ability to predetermine behavior. In 1957 Spain objected to French plans to divert water from a transboundary water system (Lake Lamoux) to support a hydroelectric project in France. In a compromise decision, an appointed tribunal ruled that Spain did not have to be notified of the project, but that France did not have exclusive jurisdiction over the resource. This helped set a precedent for what is commonly known as the "equitable utilization principle", a process or technique whose success is measured by the overall fairness of the result, rather than the particular criteria used to make resource-use decisions. [It] requires states simply to behave reasonably in their use of shared waters, with reasonableness defined with reference to the existing and, in some formulations, future pattern uses of the waterway by basin states.[10]

A recent International Law Commission Report cites two other major conventions with regard to the "non-navigational uses of international watercourses. One is the prevention of significant harm to other states, also known as *sic utere tuo ut alienum non laeda*: use your property in a way not to harm others. The other is the obligation to notify and inform others of any major alteration plans.[11] Indeed, Principle Nineteen of the Rio Declaration demands that prior consultation will be given "on activities that may have a significant adverse transboundary environmental effect." Lest we be swayed by these normative developments, Peter Gleick properly cautions us, reminding us that "[r]ecent events such as the destruction of the Peruca Dam in the former Yugoslavia, the controversy between Hungary and Slovakia over the Gabcikovo Dam on the Danube, and the continuing water disputes throughout the Middle East suggest the limited influence of international water law when other political interests are paramount."[12] There is also a great deal of variance amongst the scope and level of institutionalization; as we shall see, the Mekong Development Committee has a mandate that goes clearly beyond sharing the river, while most agreements concern themselves with much more specific agendas. Agreements also vary according to region. In Europe, there are four major shared river basins regulated by at least 173 agreements; in Africa, there are only 34 agreements for twelve international rivers.[13]

This essay will focus on an evolving experiment in international cooperation and environmental alteration. For some time now, politicians in the Southeast Asian area have talked of "harnessing" the energy potential of the mighty Mekong, the world's twelfth longest river and the longest of its kind in the region. After years of protracted planning, it would

---

[10] A. Springer, "International Environmental Law After Rio: The Continuing Search for Equity", *Ethics and International Affairs*, 7 (1993), 115-130, 118-119. See also A.P. Lester, "River Pollution in International Law", *American Journal of International Law*, 57 (1963), 828-53; and Experts Group on Environmental Law of the World Commission on Environment and Development, *Environmental Protection and Sustainable Development* (London: Graham and Totman, 1987), 72-75.

[11] UN International Law Commission, *Law of the Non-Navigational Uses of international Watercourses*, Report of the ILC on the Work of its 43rd Session (New York: UN, 1991).

[12] Gleick, *op. cit.*, 83n6.

[13] S. Baltscheffsky, "Water Conflicts Set Nation Against Nation", *ENVIRO*, 15 (July 1993), 31-32.

seem that large related development projects, based mainly on the financial assistance of the Asian Development Bank (ADB), will proceed. Two central factors are used to justify the alteration of the river to expand hydroelectric output, increase navigational capacity, develop an extensive network of railways and roads and bridges around and across the river, and boost tourism. Firstly, this will facilitate industrial development throughout the region; and secondly, this cooperation, which presently includes that of what is perceived as the region's most threatening power, China, will facilitate greater local inter-state and intra-state stability.

The merits of these arguments must be weighed not only on their own terms but also against the social and ecological consequences of the proposed and extensive environmental alteration. As we shall see, the potential for ecological disaster is quite frightening, indicated by other experiences with large hydroelectric projects and river engineering (recent events along the Mississippi River in the United States included), and the social dislocation which accompanies forced resettlement is an important human rights issue. Given the constraining nature of the political systems in the area, there exists considerable resistence to Mekong alteration amongst riparian peoples. Non-governmental organizations are involved in protesting, and some international donor agencies are beginning to adopt sustainable development strategies. While the alteration may be mitigated by such factors, it will to some extent proceed. This is due to the twin considerations of perceived expanding energy needs in the Mekong countries, and donor countries' (especially Japan's) desire to see a benign China become engaged in the region's affairs. It then becomes imperative that environmental assessment, such as that carried out by the UNDP-based Mekong Committee and Secretariat, is well structured, vigilant, empowered, and humane. This is unlikely, given the Committee's mandate to promote economic development and the limited autonomy of such international organizations. However, research efforts such as the partly Canadian-sponsored Mekong Development Research Network may offer a glimmer of hope and, even, non-governmental organization participation.

Will the Mekong Basin become a paradigm case of a resource domain starting or exacerbating violent conflict between nation-states, or between sub-state groups? Or is there hope in the naturally occurring mutual dependency an international river represents, illustrated by the evolution of the Mekong Committee and cemented by recent expansive developments? As brief descriptions of their respective political histories will make clear, these countries are quite diverse, and exist in a rather unstable historical security context. Tensions remain high between the national governments, and between sub-state groups. However, all the Governments in the region are increasingly bonded by an economic imperative to exploit their resources, aided by foreign investors, a situation which may have more positive implications for official state-state relations than it does for both those adversely affected by development and the environment itself. The ultimate outcome of an unsustainable path of development may well be that the regional international organization surrounding the Mekong River will be undone by the conflict and hostility that path engenders. As well, the development projects planned have serious implications for the human rights of the occupants on the basin, a theme we will return to in some detail below. In this light the participation of non-state representatives, such as NGOS, may be instrumental.

3. THE MEKONG AND THE RIPARIAN NATION-STATES[14]

The Mekong is Southeast Asia's longest river, located in what is geographically known as the Indo-Malayian Realm, beginning in the Tibetan Himalayas and eventually spilling into the South China Sea. Over 475,000 million m^3 of water flows from the Mekong into the Sea each year. Although it has been used by civilizations long past, including the Khmer emperors who constructed an elaborate irrigation system in the Angkor area in the 9th century, the river remains relatively untouched except for major hydroelectric complexes in China, Thailand and Laos. The lack of alteration may be ascribed to the lack of security in the area, despite an immense amount of effort devoted to securing a conducive diplomatic context. Now that the regional interstate military threat has diminished, obstacles to large-scale change come mainly from lack of funding in some parts of Southeast Asia and domestic opposition in others.

We should not underestimate the visionary appeal of the river's potential: the Mekong Committee's first Executive Agent wrote that the "Lower Mekong has thus [far] been a giant asleep -- a source of tremendous potentialities for power production, irrigation, navigation, and flood

[14] An excellent source, used here, for short synopses of political histories can be found in J. Krieger, Editor in Chief, *The Oxford Companion to Politics of the World* (New York: Oxford, 1993). Biodiversity and conservation aspects may be found in S. Braatz, "Conserving Biological Diversity: A Strategy for Protected Areas in the Asia-Pacific Region", *World Bank Technical Paper* No. 193, Asia Technical Development Series (Washington: World Bank, 1992).

control, but a source virtually unutilized."[15] This faith in panacea-like consequences colours all the official documents which justify further research into Mekong alteration, but the local government's continued interest in the Committee as well as sustained contributions from donor nations shows this to be a widely held optimism. Prince Wan Waithayakon, the Foreign Minister of Thailand and former President of the United Nations General Assembly, told a dinner crowd honouring the Americans who had done some initial research on the area in the late 1960s, "We hope that the Mekong Project will provide a shortcut to industrialization in this region."[16] And it is little wonder: the Pa Mong Dam, if constructed on the Thai-Laos border as originally planned, would have provided double the annual marketable energy produced by the Aswan Dam in Egypt, and three to four times that of the Hoover Dam in the United States. (It would also have displaced 300,000 to 400,000 people, but we will return to that later.)

Before briefly discussing individually the six nations involved, we should note that some 50 million people live in the Lower Mekong Basin Area alone, many of them relying on river water for small-scale agriculture, cross-river trade, and fish. Amongst these people lies a wealth of practical agricultural and aquacultural knowledge, and an awareness of the requirements of sustainable development. Unless one accepts the rather lame argument that Mekong alteration will benefit those displaced by it in the long run, we must conclude that these are the people being asked to pay the highest price for the industrial path chosen by Mekong area states. We will expand on this theme later in the paper, but suffice it to say that for the riparian peoples, their habitat itself is at stake. They may be citizens of different countries, but they are linked by this ecological similarity, and are those that will be directly affected by any future efforts to alter the Mekong River Basin environment. Short introductions to the nation-states in which they live follow.

---

[15] C. Hart Schaaf and Russell H. Fifield, *The Lower Mekong: Challenge to Cooperation in Southeast Asia* (Toronto: Van Nostrand, 1963), 74. Schaaf was appointed Executive Agent in 1958.

[16] *Ibid.*, p102.

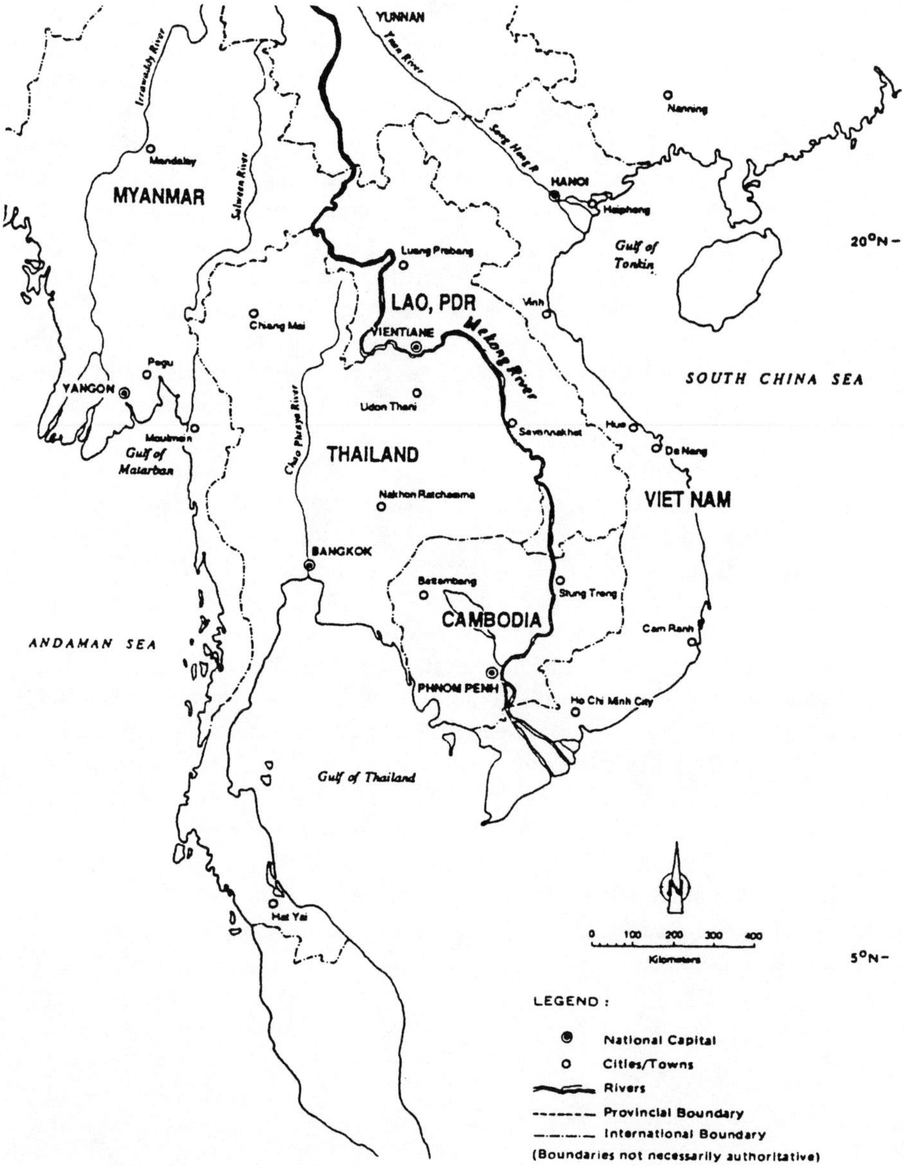

Figure 6.2: Map of the Mekong Basin

Source: map courtesy of the Royal Thai Embassy, Ottawa, Canada.

I. CHINA

The Mekong begins its arduous journey in the Tibetan highlands. In 1950 Tibet was annexed by the People's Republic of China, and since 1965 has been known as the Tibet Autonomous Region (TAR). It is the more southernly Yunnan province, however, that is most interested in the Mekong. China is one of the more prolific dam builders in the world today, exemplified by contemporary plans to create the world's biggest dam along the Yangtze, the Three Gorges project.[17] Along the Mekong, the 1250 megawatt Manwan Dam in Yunnan is to be joined by many others. Yunnan officials want to derive power from the river (mainly for Kunming City) as well as the ability to transport mineral exports, in particular phosphates, south along the Mekong to reach world markets. The latter would be made possible by dynamiting the rapids along the stretch of river that divides Burma and Laos.[18]

China's involvement in Mekong politics is of recent origin, the result of twin efforts: an attempt to open up Yunnan province to the world economy and an attempt to further China's diplomatic presence in the region (China is already the principal supplier of aid and weapons to the military government in Myanmar). It seems now that China will become an official member of the Mekong Committee, and Yunnan is already a member of the Mekong Development Research Network sponsored partially by Canada's International Development Research Centre.[19]

---

[17] For a critical evaluation of the Three Gorges project see Margaret Barber and Grainne Ryder, ed., *Damming the Three Gorges: What Dam Builders Don't Want You to Know* (A Critique of the Three Gorges Water Control Feasibility Study), Second Edition (Toronto: Probe International and Earthscan, 1993). I have reviewed this book in *Earthkeeper: Canada's Environmental Magazine*, December 1993, 35-36.

[18] One author claims that today "China is to play a part in making the Mekong River navigable for nearly all its length, carrying more industrial traffic that the Rhine by the year 2000." Depending on one's perspective regarding environmental security implications, this is either too optimistic or too pessimistic an estimate. But there is little doubt that an expanding Chinese economy will have immediate effects on the Mekong. W. Hutton, "Chinese Giant Gets Ready to Shake the World", *The Globe and Mail*, Dec. 30, 1993, A21.

[19] I am grateful to Jingjai Hanchanlash, Regional Director, Office for Southeast and East Asia, IDRC, for sending me information on the Mekong Development Research Network, based at Chulalongkorn University, Bangkok, Thailand, in September 1993. The Mekong Development Research Network (MDRN) was established in Bangkok in 1991. "Its primary objective is to provide a new forum for cooperative effort among the Mekong countries so that water resources can be better utilized and the Mekong ecosystem better preserved to improve the quality of life for the peoples along its banks." See C. Wheeler, "The Mekong: From Tibet to the South China Sea", *IDRC Reports* 21: 1 (April 1993), 26-27. The Canadian International

It is important to understand the geopolitical context of China's newfound interest in the Mekong. Quite frankly, a heavily-armed China scares policy makers both in Indochina and abroad, and its recent September 29, 1993 speech at the United Nations and subsequent underground testing of an 80 to 90 kiloton nuclear device did little to relieve such fears. Its involvement in Mekong alteration will provoke the scrutiny of others, such as the United States and, in particular, Japan, the largest shareholder of the Asian Development Bank. Japan "wants to see stronger links between Burma, Indochina, and the ASEAN nations as a way of resisting potential Chinese excesses." At the same time, if China's involvement is perceived as part of a joint partnership, China seems less threatening to its neighbours and the Japanese. "Helping Yunnan to integrate economically with its neighbours will give China a vested interest in not behaving aggressively ... The alternatives, in any case, might be futile."[20]

II. BURMA

The Mekong next graces the edge of Burma, another non-Committee member which has become increasingly involved in Mekong affairs. There are over 40 million people in contemporary Myanmar, the majority of Burmese origin and Buddhist religion. (Many persecuted Rohingya Moslems have fled into Bangladesh, and many others have fled Burma for Thailand.) A former British colony, Burma was governed since 1948 under the banner of socialism and then, after the collapse of the Burma Socialist Party in 1988, a military junta, the State Law and Order Restoration Council (SLORC) which seized power in September of 1988. This junta has subsequently refused to accept the results of a 1990 election. Though somewhat of an international pariah, the SLORC regime has nonetheless engaged in numerous joint ventures with foreign investors, including Thai, Chinese, Korean, and Japanese. These investments are, typically, resource-extraction-oriented with a strong emphasis on forestry. SLORC's environmental and human rights record has been criticized by activists such as Desmond Tutu, who has written of the "wanton destruc-

---

Development Research Centre was instrumental in bringing together the nations involved, a theme we will return to later.

[20] Both quotes from P. Handley, "River of Promise", *Far Eastern Economic Review*, 16 Sept 1993, 68-72, 70.

tion of Burma's rainforests and fisheries, as well as reported links to the immensely lucrative heroin trade."[21]

The main South Asian concern with Burma may not be the human rights of its Moslems, but the increasing docking rights gained by the Chinese army in the bay of Bengal. Japanese defense experts worry about Chinese presence near Mergui, which is 400km north of Penang (Malaysia) and the entrance to the Strait of Malacca, through which a large portion of Japan's oil imports must pass. India, despite cooling relations with China, continues to question Chinese actions and motives in the area as well, including a Chinese radar installation on the Coco Islands. One analyst purports that the total arms sales delivered or on order from Yunnan to Rangoon is in excess of $1.2 billion; this includes six Hainan class fast attack vessels in 1991, and four more in 1993, "accompanied by 70 Chinese naval personnel -- over half of whom are middle-rank officers."[22]

Burma's involvement has, overall, been minimal. A Global Environmental Facility (administered by the World Bank) United Nations Development Program (UNDP) project of 3.9 million American dollars will focus on the sustainable development of all its natural resources. Rangoon's increasing dependence on China for arms and supplies may, in the long run, be superseded by such outside aid, though this is unlikely if the SLORC regime remains in power. But as a riparian state, Burma has the international legal right to participate in serious alterations along the Mekong. The Thais have expressed an interest in getting Burma to officially join the Mekong Committee. Meanwhile, the international community struggles with the prospect of lending the SLORC legitimacy.

III. LAOS

Laos is the smallest of the nation-states involved in this study, with barely four million predominantly Buddhist people. It and Cambodia gained independence from France in 1953. The Pathet Lao ("Marxist-Leninist" Party) won a protracted civil war with the Royal Lao Government, and the Lao People's Democratic Republic (LPDR) was proclaimed in December of 1975. An August 1991 Constitution has but enhanced the powers of Party Chairman Kaysone Phomvihane, who is also state presi-

[21] "Burma as South Africa", *Far Eastern Economic Review*, 16 September 1993, page 23.

[22] R. Tasker, "Last Refuge", *Far Eastern Economic Review*, 16 Sept. 1993, 27.

dent. Major opposition to the ruling regime can be found amongst the opium-growing Hmong (Meo), who constitute some ten percent of the population. Laos, along with Vietnam and post-1979 Cambodia, has formed the "solidarity bloc", a socialist antithesis to ASEAN capitalism. This was facilitated by foreign assistance from the now defunct USSR. However, much like the other states and despite western trade embargoes, after 1988 Laos has begun seeking foreign investment, and the Thais have responded favourably, especially in timber extraction. Despite heavy bombardment of the "Ho Chi Minh Trail" by the United States, Laos still retains large tracts of undisturbed rainforest, and will no doubt be the site of future international environmental concern as rainforest dwindles quicker elsewhere.[23] Logging rights are monopolized in many areas by a powerful military-owned company, BPKP.

Despite its small size, land-locked Laos is a very important part of the Mekong situation: it contains 50% of the Lower Mekong's hydroelectric potential, and is relatively thirsty for water since only 1% of Laos' 800,000 hectares of arable land are irrigated. Over 95 percent of Laos's total electricity generation is from hydroelectric production.[24] Two thirds of Laos' foreign trade crosses the Mekong - to and from Thailand - every year, and a great deal of illicit trade is known to occur. (Australia is currently funding the construction of a bridge between Thailand and Laos.) The 150 megawatt Nam Ngum Dam, on a Mekong tributary, provides a third of Laos' foreign exchange through power exports to Thailand. Though other hydropower facilities exist, in June of 1993 energy-hungry Thailand signed a memorandum of understanding to develop more hydroelectric dams in Laos.[25] Laos has also recently signed a friendship treaty, and a transport agreement, with its northern neighbour China.[26]

---

[23] See R. Stevens, *The Trail: A History of the Ho Chi Minh Trail and the Role of Nature in the War in Vietnam* (New York: Garland, 1993). For earlier accounts of eco-destruction during the war, see A. Westing, "The Cratering of Indochina", *Scientific American,* May 1972, 21-29; J. Neilands, et. al., *Harvest of Death: Chemical Warfare in Vietnam and Cambodia* (New York: Free Press, 1972); and J. Schell, *The Military Half: An Account of Destruction in Quang Ngai and Quang Tin* (New York: Knopf, 1968). See also the section devoted to "Military Conflict and the Environment" in Chapter Three of this study.

[24] UN Statistical Office, *Energy Statistics Yearbook 1987* (New York: UN, 1989). By contrast, Canada relied on hydroelectric power for 63.7% of its energy in 1987.

[25] This was signed during Prime Minister Chuan Leekpai's June visit to Vientiane. Two sites are currently being studied. *Far Eastern Economic Review,* 17 June 1993, page 67.

[26] *Far East Economic Review,* 16 Sept. 1993, 15.

IV. THAILAND

Thailand has been the foremost and most vocal proponent of major development plans along the Mekong. It stands alone in the region as the only "Little Asian Dragon", with astounding double-digit (over ten percent) economic percent growth in the late 1980s (the highest of any country anywhere), facilitated by rapidly expanding industrial and agricultural export sectors. It is also the only historically non-colonized state, with at present some 60 million citizens. Thailand has been subjected to military rule on an off-on basis for some time. Civilians and military officers overthrew the Monarchy in 1932, and as recently as 1991 there was a coup unseating conservative leader Chatichai Choonhavan, elected in 1988 (in May 1992, civilians demonstrated against the coup, and forced yet another return to fragile civilian rule). Throughout, the Thai monarch has remained the national symbol and leader of the Buddhist religion. A Chinese-Thai minority has dominated the economy, and is becoming more involved in politics. The Thai government has conflictual relations with indigenous hill tribes such as the Hmong and the Karen. It also has ongoing disputes with Malaysia over ocean fishing rights.[27] The Royal Thai Government declared a 20-point National Forest Policy in 1985; after a severe storm and landslides in 1989, the government imposed a total ban on logging (deforestation is one of the chief causes of landslides.) A proposal by the Thai Interior Ministry to reopen closed borders with Cambodia has raised great concern in 1993 because of the persistence of the Khmer Rouge, but the demand for raw timber is increasing.

Thailand is indeed a growing economy. Electricity requirements have been officially projected to double between 1988 and 1996 to 11,066 megawatts. The principal justification for damming the Mekong would be to supply this electricity, but Thailand is interested in a variety of other prospects. It wants to use the Mekong to irrigate the arid northeastern province, Isan, half of the country's arable land. Thailand's export-led development has been partly dependent on increases in commercial agricultural expansion in the countryside, and Isan is the least developed area in this context. The plans for Mekong diversion, which have raised great concern in downstream Vietnam, are known as the Kong-Chi-Mun and Kok-Ing-Yom-Nan projects. And the river must, Thailand feels, be made

[27] See Mark J. Valencia, "International Conflict Over Marine Resources in South-East Asia: Trends in Politicization and Militarization", in L.T. Ghee and M.J. Valencia, ed., *Conflict Over Natural Resources in South-East Asia and the Pacific* (Singapore: Oxford and United Nations University, 1990), 94-144, 116.

safe for tourism. Indeed, supplying leisurely trips up the Mekong for visiting Asians would be an enormous tourism opportunity. The demand for electricity remains central, however, and Thailand has plans to expand its nuclear sector as well. The threat of this nuclear growth is often used as an argument in favor of Mekong alteration. Meanwhile, Thailand's record for industrial and environmental safety is miserable; the nation seems to have some predilection for disaster.[28]

Thailand clearly sees itself as the economic leader of the region, and as the driving force behind increased cooperation in the name of capitalist competition. Thailand officials see the Chinese threat not so much in military terms but in economic potential: the thought of cheap Chinese goods flooding Thai markets via Mekong transportation is not a pleasant one. But they are more than ready to take this risk, and there is an atmosphere of political optimism regarding Mekong development plans.[29] It is clear, however, that Thailand does not want a great deal of interference from international organizations like the Mekong Committee, which in fact has its headquarters in Bangkok. A recent dispute with the Executive Agent, Chuck Lankester, may have resulted in his eventual reassignment to UNDP offices in New York.[30]

V. CAMBODIA

After touching all the above states, the Mekong runs through the heart of Cambodia, which contains the vital Tonle Sap and Great Lake, also known as the Mekong's "natural regulator" (the Lake stores parts of the peak flow from July to December, and releases it over winter.) The modern history of Cambodia is of course a troubled one as well. France,

---

[28] A look at recent events suggests this much. September 1990: 91 people burned to death when a lorry carrying gas cylinders turned over; February 1991: 171 people killed after they gathered around an overturned dynamite lorry which exploded; March 1992: 119 Buddhist pilgrims drowned when a ferry they were taking to an offshore shrine collided with an oil tanker; and 9000 tons of molasses leak from a sugar mill into the Nam Pong River, killing uncountable fish and river wildlife; May 1993: 189 people, mostly women workers, died as a toy factory with its emergency exit doors locked burned; August 1993: over 100 died as the Royal Plaza Hotel, onto which four stories had been illegally added, collapsed in Nakhon Ratchasima. Source: *The Economist*, August 21, 1993, 30.

[29] I am grateful for discussion and information to Annop Soonjan, Foreign Officer, Royal Thai Embassy, Ottawa, Canada.

[30] This is related to the dispute with Vietnam, mentioned below. Lankester was, it should be noted, very pro-alteration; it would seem the conflict was over jurisdiction rather than ideology. Lankester has not made himself available for comment.

deep in a losing war with Viet Nam, granted King Sihanouk's government of Cambodia independence in 1953. Prince Sihanouk's "Royal Buddhist Socialism" ruled Cambodia from the Geneva Accord-inspired elections of 1955 through to 1970. General Lon Nol assumed the presidency of a new Khmer Republic in 1972. In April of 1975 the Khmer Rouge, led by the infamous Pol Pot, seized power, established the Government of Democratic Kampuchea, and quickly set about engaging on a course toward human, and ecological, carnage. Secluded Kampuchea pulled out of the Mekong Committee at this time. In January of 1979, Vietnam successfully invaded Kampuchea (which, in 1989, was again renamed Cambodia). The Vietnamese-installed government was, initially, "communist" in character, but a 1991 decision to privatise state enterprises and decollectivize agriculture outdated this label. In opposition to the Vietnamese puppet government, the "Coalition Government of Democratic Kampuchea (including the Khmer Rouge, Khmer Republic of Lon Nol, and Sihanouk's Kingdom), based itself on the Thai border and received support from China, the United States, and ASEAN. In recent developments, an October 1991 Peace Agreement called for the June 1993 election which involved substantial UN peacekeeping participation, some 300,000 refugees have returned, and the monarchy is being restored in Cambodia.

Cambodia's itinerary Supreme National Council adopted a complete ban on the export of Cambodian logs in 1993, which was subsequently endorsed by the UN. Thai companies which have harvested logs there before the imposition of the ban have demanded the Thai government defy the ban, and there is little doubt that a great deal of trade between the Khmer Rouge and Thai logging concessions continues, and that this will continue so long as the Khmer Rouge remain an organized but delegitimized force heavily dependent upon export earnings.[31] This is a serious issue of dispute between the newly formed Cambodian regime and the Thais.

[31] The initial Cambodian ban was imposed for three apparent reasons: to save a rapidly depleting forest base; to prevent Cambodian political parties from using revenue from lumber sales to influence the 1993 election; and to punish the Khmer Rouge, which by then had dropped out of peaceful negotiations and was benefiting from trade with Thai concessions who had turned to Cambodian products after the Thailand ban on logging of 1989. See P. Shenon, "Thailand Stirs Criticism Over Cambodian Timber", *The New York Times* May 2, 1993, page 21. Meanwhile, Khmer Rouge exports include not only timber but precious gems as well, much from the gem-rich Cardamom mountains in southwestern Cambodia, where it grants access to Thai miners. Z. Abuza, "The Khmer Rouge Quest for Economic Independence", *Asian Survey: A Monthly Review of Contemporary Asian Affairs*, 33:10 (1993), 1010-1021.

Cambodia is still in the genesis of what will hopefully be a more stable future, but future tension over areas like timber trade will raise inevitable difficulties. Due mainly to the continued onslaught of war, Cambodia's irrigated farmland has fallen by 80% in the past 20 years, and its cities still rely on fuel driven generators for power. There are obvious needs for utilizing whatever natural resources can be used in the area. Cambodia is at present seen as the younger sibling in the region, still unsure of itself. It was officially reintegrated into the Mekong Commission in early 1993. Part of the difficulty involved in reinstating Cambodia had little to do with Cambodia: according to the original 1957 Mekong Committee rules, all downstream states would have a veto over large-scale alteration plans. Since Thailand believed the Vietnamese would veto its irrigation diversion projects in Isan, it was against allowing Cambodia back in since this would, in effect, reinstate the original arrangements. In February of 1993 in Hanoi, Vietnamese representatives conceded Thailand's right to diversion projects, pending upcoming agreements about future development rules, and Cambodia once again became part of the official Mekong Secretariat decision-making process. This was heralded as a great step towards greater stability in the area.

VI. VIETNAM

The river finally stretches into the often disputed territory of the Socialist Republic of Vietnam, which at 66 million (1992) is the most heavily populated country in southeast Asia. Most of this population is concentrated in two rice-growing areas, the Red River Delta in the north and the Mekong River Delta in the south. Vietnam is a tired war veteran. In August of 1945 Ho Chi Minh's Vietnam Communist Party (VCP) seized power and a protracted, eight-year war with former colonial power France followed. In 1954 Vietnam was partitioned, the VCP ruling the north with large-scale Chinese and Soviet aid. In 1960 the North increased its efforts to aid the communists in the civil war in the south, and eventually the United States became actively, and seemingly intractably, involved. The war with the U.S. ended officially in 1975 with the integration of the south; large numbers of ethnic Chinese and southern Vietnamese have fled the country by boat or land into southern China or elsewhere. Natural disaster in 1977 and 1978 added to the additional turmoil of war with the Khmer Rouge in late 1978 and China in early 1979. Forest cover was reduced by as much as 50 percent during the war years, attacked by a combination of chemical defoliants, bulldozers, napalm, and saturation bombing.

Western aid and trade embargoes have led to total dependence on now non-existent Soviet aid. Regular forces were withdrawn from Cambodia in 1989; but the regime continues to clamp down on dissidence and relations with the West and China remain, to say the least, strained (an American trade embargo, to the delight of Pepsico and other American multinationals, was lifted in early 1994).

Vietnam's main interest in the Mekong stems from the river's immense contribution to its agriculture, which is booming: between 1980 and 1990 Vietnam doubled its rice production to about 10 million tonnes, and became the world's third largest rice exporter. Plans have been made to use dams and dykes to keep out the annual intrusion of sea water from the South China Sea, despite the potential of rice-shrimp culture and the acidic dangers associated with pyrite exposure. Even more controversial, upriver diversion of the Mekong, such as would result from intensive damming along the Thai-Laos border or from Thailand's plans to irrigate Isan, could harm crops all along Vietnam's share of the river. Indeed, divisions between Thai plans for alteration and Vietnam's concerns over its effects have been vocalized in the recent past, despite official pronouncements of mutual commitment. Vietnam, which of course has nowhere near the industrialization realized by Thailand, is less concerned with the hydroelectric potential of the river. Vietnam contains over 2,200 rivers, including the Da, Sesam, and Dong Nai, which gives it a very large theoretical hydropower potential, if but a small installed capacity exists at present. It is suspected that there are large petroleum deposits in the country, and in the Mekong Delta as well, but these are largely unexplored.[32] The Mekong Secretariat, the World Bank and a Dutch consortium are preparing a long-term development agenda for the Delta area.

Sino-Vietnamese relations, which have oscillated in the past but remained largely acrimonious, are not at a high point. Beyond seemingly perpetual border disputes, it is well known that there are significant oil and natural gas deposits in the South China Sea area, particularly in the Paracel and Spratly Island groups. The Chinese military has a landing strip on Woody Island, the largest of the Paracels. Last year China unilaterally "awarded exploration rights for a disputed part of the sea -- an area

---

[32] Nguyen Huu Ninh, et. al., "Energy in Vietnam", UN Centre for Science and Technology for Development, Advanced Technology Assessment System, No. 6, *Energy Systems, Environment and Development* (New York: 1991), 37-38.

that Vietnam insists is part of its continental shelf -- to Crestone Energy Corporation, an American oil company."[33]

In summary, then, the purposeful alteration of the Mekong is an obviously enticing prospect for the managers of industrialism in the area, in order to provide electricity and river transportation, and it could contribute to increased agricultural yields. Some alteration of the Mekong area and the river itself is inevitable. The Mekong will serve as the central point in an intensive region-wide development plan, sponsored mainly by the Asian Development Bank. A conference in Manila in August 1993 produced a list of planned railways and roads, and feasibility studies have begun on four road projects. Most of these projects involve upgrading existing roads, such as the one between Dali in Yunnan and Lashio in Burma, and the road linking Bangkok, Phnom Penh, and Ho Chi Minh City. Further study is planned for a major cross-border power grid; Thailand would be the market, Laos, Burma and China the main suppliers. Although the Asian Development Bank has begun to play a major role in the region, most alteration proceeds through the channels established by the Mekong Committee. The volume of development assistance allocated to the Mekong Secretariat has been impressive, and it has been demonstratively shifting toward support for "Basinwide", or international, projects.[34] It is to that organization we now turn.

---

[33] N. Kristof, "The Rise of China", *Foreign Affairs* 72:5 (1993), 59-74, 64. Kristof reports: "[t]he head of the company told me he was promised the full backing of the Chinese navy in exploring the area."

[34] Total cash and in-kind contributions, in millions of U.S. dollars: 1988, 9.422; 1989, 7.367; 1990, 10.271; 1991, 8.958; 1992, 11.794; total, 1988-1992, 47.811. Programme support by source: European Community, 31.8%; Sweden, 17.9%; UN Agencies, 12.9%; Australia, 12.3%; Switzerland, 10.5%; Finland, 10.5%; Japan, 2.2%; others, 1.9%. In 1987, 89% of Secretariat projects were national-based, and only 11% basinwide; the ratio was 85 - 15 in 1988; 75 - 25 in 1989; 61 -39 in 1990; 51 -49 in 1991; and 52 - 48 in 1992. (We should note that this assistance does not include the future plans of the ADB.) Statistics from Mekong Secretariat, *Annual Report 1992* (Bangkok, 1993), pages 32-33. I am grateful to Piyanart Supaphon, Documentalist, Mekong Documentation Centre, Mekong Secretariat, Bangkok, Thailand, for communication in August 1993.

4. Regional Organization and the Mekong

International cooperation has often centered around the river and its economic potential. In 1926 Thailand and then-colonist France agreed that neither would impede navigation on the Mekong; in 1949 Cambodia, France, Laos and Vietnam signed the *Convention on Maritime and Inland Navigation on the Mekong and on Inland Water Routes to Saigon.* In 1951 Cambodia, Laos, Thailand and Vietnam signed the *Statute of the Committee for Co-ordination of Investigations of the Lower Mekong Basin.*[35] Several reports, financed by the United Nations' Economic Commission for Asia and the Far East (ECAFE) and the United States, surfaced during the 1950s in support of the idea that much more research should be conducted with an aim towards establishing a permanent commission to facilitate exploiting the river's potential.[36] The resultant Mekong River Basin Project included the Committee for Investigation of the Lower Mekong, initiated under the auspices of ECAFE, in Bangkok in 1957. That year the French donated substantial hydrological and meteorological equipment to the Committee, and a research mission headed by a former Chief Engineer of the US Army Corps, Raymond Wheeler, provided the Committee with a working agenda, one to be frequently interrupted by political developments and yet never quite shelved.

Of course, the preceding narrative of the Mekong Committee's genesis is free of political context; contrary to what some might say about the ability of technical experts or even what international regime analysts label 'epistemic communities' to override political considerations, the Mekong has always been heavily influenced by such things as superpower interests and ideology. This being said, the claim (advanced by both proponents and opponents of large-scale alteration) that the river can have an almost pacifist effect on inter-regional conflict is not without some justification. Schaaf mentions the Thai-Cambodian temple border dispute of 1958: Ambassadors had been recalled, borders closed, and Phnom Penh-

---

[35] Text is in *UN Legislative Series, Legislative Texts and Treaty Provisions Concerning the Utilization of International Rivers for Other Purposes Than Navigation*, CST/LEG/SER.B/12; UN Publication Sales No. 63.V.4, 267-270.

[36] ECAFE's Bureau of Flood Control (later the Bureau of Flood Control and Water Resources Development), "Preliminary Report on Technical Problems Relating to Flood Control and Water Resources, Development of the Mekong -- an International River", 1952; Bureau of Reclamation of the U.S. Department of the Interior, "Reconnaissance Report -- Lower Mekong River Basin", 1956; ECAFE, "Development of Water Resources in the Lower Mekong Basin", 1957.

Bangkok air routes closed. Nonetheless, in the midst of this acrimony, and with the help of an American plane and diplomatic pressure, the Cambodians made it to Bangkok for what was a crucial Mekong Committee meeting.[37] This aura of cooperation despite conflict continues to this day, and now involves even Burma and China. However, the Committee's early impetus came from the American superpower that wanted to stop communism's spread in the region at all costs, and subsequent development plans, funded by the Colombo Plan and United Nations, were designed in this ideological context.

The Mekong Committee, which is officially part of the United Nations Development Programme, meets twice a year in the country currently holding the chairmanship (this is rotated in an alphabetical order). Though originally designed as a unit-veto structure, Thailand has recently insisted that these 1957 rules are obsolete in the present circumstances. The basic objective of the Committee is in fact anything but basic:

The comprehensive development of the water resources of the lower Mekong basin, including mainstream and tributaries, in respect of hydroelectric power development, irrigation, flood control, drainage, navigation improvement, watershed management, water supply and related developments for the benefit of all the people of the basin, without distinction as to nationality, religion or politics.[38] Though a 1970 "Indicative Plan to the Year 2000" was shelved due to conflict in the area, it was revised in 1987 (now entitled "Prospects for Mekong Development"), and contains four main elements: a long-term perspective of the basin's development potential; an investment plan covering 1988-2000; a concurrent programme of studies and investigations; and proposals to give the Mekong Secretariat appropriate autonomy in order to carry out its work. This autonomy is the source of tension between the Secretariat and the nation-states involved, and it is not common in international relations to give an international organization a great deal of decision-making power. The autonomy is most likely perceived as the price to be paid for the impressive amount of financial assistance that has flowed into the region under the auspices of Mekong Secretariat control.

---

[37] Schaaf, *op. cit.*, 125-7. Schaaf, writing in 1963, concluded that "the Mekong Scheme has not been involved in the global politics of the Cold War." *Ibid.*, 129. This was debatable then, and certainly seems premature now.

[38] P. Savasdibutr, "The Development of the Lower Mekong River Basin", UN Department of Technical Co-operation for Development, *River and Lake Basin Development*, Natural Resources Water Series No. 20 (NY: UN, 1990), 170-179, 176.

There can no doubt that the Mekong Committee was and is directly concerned with the environmental consequences of river alteration, and it may even be one of the more advanced examples of an international environmental assessment mechanism.[39] Indeed, environmental attention has reached an organizational peak: in 1991 the Secretariat created an Environmental Unit within its Technical Support Division, and there has been an important Water Quality Monitoring Network in place since 1985, funded by Sweden. The question is whether the Secretariat will be able, in the long run, to enforce environmental guidelines among the participating states, and given the limited autonomy of UN-based international agencies, it would be chimerical to expect it to do so.

5. GES and International River Alteration

As discussed earlier in this book, global environmental security is offered here as unifying theme, one which pulls together the linked concepts of ecosystem health, decision-making and risk aversion, and human rights, and the changing but resilient nature of national sovereignty. We might briefly recall what was implied by the term GES. It is essential to have some scientific understanding of ecology, or of what we have termed the *basic health needs of ecosystems.* (There is room for relativity here, or for different understandings of ecology within the universal imperative of biospheric health.) In a broader sense, environmental security refers to the protection of human habitat, and efforts to avoid risking that protection such as official environmental impact assessment and citizen's opposition to destructive activities, as well as international agreements to limit pollution and promote sustainable development. The normative dimension of human-nature ideology therefore plays an essential role, since it is a certain ethical understanding of the uses and rights of nature which lead to humankind's attempts to dominate it with technology. Finally, environmental security must seen as an essential human right. Indeed, most people would agree that nothing could be more essential than the humanitarian right to a survivable environment; how this is achieved is, of course, the subject of debates related to ecosystem health, risk aversion, and human-nature relations.

---

[39] See in particular V.B. Pantulu, "A Case-Study of the Pa Mong Project: Environmental Aspects", in UNEP, *Environment and Development in Asia and the Pacific: Experience and Prospects*, UNEP Reports and Proceedings Series No. 6, 1982, 145-159.

Environmental security must be seen as a global *problematique*, an imperative strengthened by the current dictates of commercial globalization. The 1992 Earth Summit (UNCED), though predictably disappointing to environmentalists, established a precedent of dealing with biospheric environmental problems on a global basis, and of linking them with North-South relations. But we must also emphasize the local dimension of the problems associated with social living conditions within smaller communities, and the social stratification within these areas. Furthermore, environmental security as a human right must be seen on two levels, the individual and the collective. An analytic decision to relegate one of these levels to the obscure is unnecessary, providing one is aware of the inequality mentioned above. In other words, environmental security will always be discriminate, more available to some than others; yet ultimately it is needed by us all and could serve as the foundation of a conception of individual human rights that promotes sustainability. Whether it can promote equality in the process is, of course, an even more difficult question.

The development of large-scale dams for hydro-electrical and irrigation purposes involves many environmental aspects, physical, geochemical and biogeochemical, and biological. Few would deny the potential for ecological disaster associated with the development of large dam systems and, more generally, the alteration of natural river paths and cycles.[40] International rivers defy the neatness of self-contained national ecosystems, and force us to consider the indirect effects of alteration on downstream states. (And upstream states, since rivers are rejuvenated from lower lakes and oceans, and aquifers often reach two or more upland states.) These effects include agricultural runoff and pesticides, increased flooding from forestry practises and the filling of wetlands, sediment and nutrients from animal husbandry, reduced ecological capacity, increased siltation by damming, and added pollutants from industrialization and urbanization. It is generally accepted that the river's annual flooding is a vital ecological function and not, as the initial U.S. Bureau of Reclamation assumed, something that needs to be controlled; even the yearly invasion of sea water in the Mekong Delta has been used

[40] For what has become a standard work, see N. Hildyard and E. Goldsmith, *The Social and Environmental Effects of Large Dams*, Volume One and Two (Cornwall: Wadebridge Ecological Centre, 1984 and 1986). Recent events in the American Midwest support claims that river alteration eventually leads to river reclamation. See William Stevens, "Why the Waters Must Flow Free", *The Globe and Mail*, July 21, 1993, A9.

profitably by innovative Vietnamese farmers, who know a thing or two about acidity and salinity.

International concern has centred on those who most often pay the full price for environmental alteration, and end up displaced by a process which, ostensibly, is designed to improve the quality of their lives. Some have argued that resettlement is not really an environmental problem. Harold Frederiksen writes for the World Bank: one "concept inconsistent with sound institutional principles is to define resettlement of people as an 'environmental' area of government activity ... incorporating the social/political oriented activities of resettlement into the already confusing 'environmental' field confounds sensible structuring of government agencies for managing its resources and its environment." The argument is specious; though it might (or might not) make sense in a policy-oriented paper, a broader analysis which defines environmental security and human rights as two sides of the same coin cannot take such a neatly divided approach.[41] In the case of river alteration, we are primarily speaking of what are often termed riparian people, though it also effects the hinterland when roads, bridges, transmission lines, and other infrastructural implements are constructed. Substantial migration accompanies the creation of anthropogenic lakes (or reservoirs), involving four main categories of people:

a) those who must relocate because their homes and fields will be partially or totally inundated by the reservoir (the relocatees)
b) those among whom the relocatees must be resettled (the hosts)
c) those lake basin inhabitants who are neither relocatees or hosts
d) immigrants who move into the lake basin and seek new opportunities that accompany dam construction and reservoir creation.[42]

This is a fundamental human rights issue, since these people are often involuntary participants in this process. This seems obvious for a) and b), yet the workers moving into these areas (d) are often there because of the environmental alteration of their local habitat as well as forced migration policies of governments. For example, workers at the giant Japanese financed Asahan aluminium project in North Sumatra,

---

[41] Frederiksen quote is from his "Water Resources Institutions: Some Principles and Practices", World Bank Technical Paper No. 191 (Washington, 1992), 19.

[42] UNEP, *The Environmental Impacts of Production and Use of Energy: An Assessment Prepared By the UNEP* (Nairobi: Tycooly, 1981), 266.

Indonesia, have been associated with that nation's controversial Transmigration Programme. The Asahan project includes a series of power plants on the Asahan river and at Lake Toba, and a smelter at Kuala Tanjung, near Medan. Eighty percent of the electricity generated by the plants is consumed by the smelter, and as of 1990 none went to the 20,000 aboriginal people living downstream, some of whom were evicted.[43] The most recent infamous case involving forced resettlement, and opposition to it at local and global levels, is that of the Indian Government's plans to build, initially with World Bank financing, the Sardar Sarovar Dam on the Narmada River. The World Bank has recently disinvolved itself from the project, prompting New Delhi to ask for a loan cancellation, but the Indian Government seems determined to push ahead with a plan that will displace from 100,000 to 250,000 people.[44] As mentioned above, a proposed project on China's Yangtze River, the Three Gorges Dam, will involve much higher displacement figures.

With regard to the link between environment and displacement, attitudes have indeed changed a great deal. As late as 1983 the Mekong Secretariat would write that, though it would cost over four billion dollars to construct, "the benefits of the proposed enormous Pa Mong Dam would be equally impressive. Thailand would get much-needed power at a reasonable price. Laos would have its foreign exchange problems solved for decades [!], Kampuchea would find its hydropower sites (such as Sambor) greatly enhanced because of flow regulation, and Viet Nam could irrigate the entire delta, some 3 million *ha,* without having to fear salinity intrusion from the sea..." Remarkably, in this 1983 report, there was no mention of resettlement problems, though construction of the Pa Mong would have flooded 3700 km^2 of farmland and forest. The Pa Mong, the Secretariat went on to say, would be but one of six or possibly even seven mainstream projects, and the entire energy potential of the

---

[43] See Yoko Kitazawa, "The Japanese Economy and South-East Asia: The Examples of the Asahan Aluminium and Kawasaki Steel Projects", in L. Ghee and M. Valencia, eds., *op. cit.*, 51-93.

[44] See in particular J.R. Wood, "India's Narmada River Dams: Sardar Sarovar Under Seize", *Asian Survey: A Monthly Review of Contemporary Asian Affairs* 33:10 (1993), 986-984. Wood concludes: "[t]he World Bank withdrawal may slow down the construction schedule, but the unfolding story of the Narmada controversy indicates that too much is at stake -- economically, legally, and politically -- for current developmental plans to be reversed" (969). See also B.D. Dhawan, ed., *Big Dams: Claims, Counter-Claims* (New Delhi: Commonwealth, 1990).

basin was assessed at the equivalent of 1 million barrels of oil per day.[45] Yet by 1991, Erik Skoglund, the Mekong Committee's senior environmentalist who recently resigned due to differences between himself and the Committee regarding its commitment to pressing ahead, said that "no donor in the world will allocate funds for Pa Mong."[46] Controversial dam projects, such as the one on Thailand's Moon River, are under construction, but it would appear as though megadams on the Mekong are less likely today than they were in the pre-war days.

Experience in the area is probably as good a warning as any. One study of Nam Pong Dam evacuees in Thailand (some in government-sponsored settlements, some not, and a 'control group' living near the reservoir), concluded that most of the evacuees (2,750 families) were clearly disadvantaged since leaving as their monetary costs were well in excess of compensation receipts, they ended up with less land and less livestock, and in general lower incomes. The "planned" settlement evacuees seemed worse off than the others.[47] Officially, the Mekong Committee still talks about the construction of the Pa Mong, renamed the "Low Pa Mong" to reflect mitigating changes. The Low Pa Mong would be 207.5 metres above sea level, as opposed to the originally proposed 250; this would supposedly lower resettlement requirements from 300,000 to 59,000 persons.[48] The Mekong Secretariat's Annual Report of 1992, how-

---

[45] Mekong Secretariat, "Cooperation in the Lower Mekong River Basin", in UN Department of Technical Cooperation for Development, *Experiences in the Development and Management of International River and Lake Basins*, Natural Resources / Water Series No. 10 (New York: UN, 1983), 245-251, 249.

[46] Quoted in M. Hiebert, "The Common Stream", *Far East Economic Review* Feb. 21, 1991, 24-28. Skoglund was on loan from the Swedish International Development Agency.

[47] R. P. Lightfoot, "Alternative Resettlement Strategies in Thailand: Lessons From Experience", in P. Gosling, ed., *Population Resettlement in the Mekong River Basin: Papers of the Third Carolina Geographical Symposium* (Chapel Hill: University of North Carolina, 1979), 28-38.

[48] It was in fact Canadian funding that allowed the Mekong Secretariat to complete and publish its Low Pa Mong Optimization Study in early 1992; the study was done by a Canadian engineering firm, Acres International, funded by the Canadian International Development Agency. The Canadian connection in Asia is an expanding one. Canadian electrical utilities and Power Corp. of Canada have joined in a $100 million dollar venture to cover a wide range of investments in Asian power production, and Hydro-Quebec is a partner in the controversial Three Gorges dam project. Barrie Mckenna, "Utilities Join Power to Tap Asian Demand", *The Globe and Mail*, October 8, 1993, B1. The pattern of development assistance-power firm collaboration is a familiar one. An international consortium, Statskraft, composed of power companies from Denmark, Finland, Norway, and Sweden, is planning to build a 210MW dam on the Theun River (a major Mekong tributary). The feasibility study and environmental impact assessment was conducted by Norpower (a Norwegian engineering firm) and sponsored by Norway's aid agency, NORAD. Some of the Laos dams planned will in fact

ever, made no claims that construction would proceed, only briefly mentioning the completion of an optimization study.

Attention drawn to the plight of the displaced as well as other environmental concerns was raised by NGOs[49] in the area, and the Pa Mong has become Southeast Asia's Sadar Soravar. Similarly, Thailand's Nam Choan Dam in western Kanchanabuei Province was stopped due to local opposition. It would have drowned much of the Thung Yai Wildlife Sanctuary, which has since been recognised as a United Nations World Heritage site. One should not underestimate the surprising influence of the environmentalist groups in Thailand, despite its rather authoritarian political climate, as writes a group of energy experts, looking into Thailand's energy production future:

> Thailand foresees a modest increase in installed hydro capacity between 1991 and 2001, but hydro's share in the capacity mix falls from 25% to 16%. The minimal increase in hydropower reflects the strength of Thailand's environmental lobby, which contributed to the thai government's recent decision to scrap plans for a major hydro project, the Nam Choan Dam [see above]...Thailand's utility recently was forced to postpone the construction of the planned Ao Phai coal plant, due to public concerns regarding potential environmental impacts.[50]

Other large dams have failed to materialize for other reasons. The planned Nam Theun 2 Dam in Laos, which would flood 340 km^2 of the Nakay Plateau and displace 4,000 people, was apparently shelved because the logging firm BPKP didn't like the idea of opening the area up to other contractors.

But this should not obscure the fact that, regardless, significant alterations will take place in the upcoming years along the Mekong. Calls for the less destructive options of energy conservation and efficiency will

---

proceed almost directly along the lines of private investment, through the "Build, operate, and transfer" (BOT) scheme, in which "private investors can accrue profits from electricity sales for years before transferring ownership to national governments." *TERRA Lower Mekong River Basin Briefing,* September 6, 1993.

[49] Special mention should be made of the Burma-Indochina NGO TERRA, or Towards Ecological Recovery and Regional Alliance. I am grateful to Grainne Ryder for sending me related information from Bangkok, and I highly recommend a new photo-essay book which she introduced, *The Mekong Currency* by L. Sluiter (Bangkok: TERRA, 1992).

[50] S. Myers, et. al., "Prospects for the Power Sector in Nine Developing Countries", *Energy Policy,* 21:11 (1993), 1123-1132, 1127-1128.

not be heard over the rumble of industrialization in a region where export oriented development, complete with Export Processing Zones which attract foreign capital, is the dominant paradigm.[51] Although large-scale damming is hardly the answer, when it comes to energy production one must, after all, consider alternatives. For example, if the Thais go ahead with large-scale nuclear facility construction, they will be endangering the region as much as any dam. It would seem as though the way of the future will be medium-sized, multi-purpose reservoirs such as those proposed along the Quae Yai, Tapi-Phum Duang, and Yom rivers in Thailand.[52] These dams will have negative effects as well, and an optimist would hold that the Mekong Secretariat remains a forum in which those effects can be discussed. The paradox -- that environmentalists must to a certain extent rely on the very institutions that promote alteration to mitigate its harmful effects -- is in the short-term inescapable. Greater transparency in the Secretariat's proceedings would help facilitate greater dialogue between the intergovernmental and substate levels on these issues.[53]

6. Conclusion

It is imperative that environmental security, defined as a fundamental human right, remains central in the future alteration of the Mekong. It could be argued, however, that this will be difficult given the nature of governance in the area, with its brutal political past and current level of citizen suppression. It is hard, in the Mekong context -- which includes for example Tibet and Burma -- to take seriously the prospects of Principle Twenty-Three of the Rio Declaration, which states that the

[51] In fairness, we must note that Thailand has instituted an energy conservation program. As Myers, et. al. write: "[i]n collaboration with an international NGO, the Electricity Generating Authority of Thailand (EGAT) prepared a Direct-Savings-Management Strategy for reducing peak demand by 225 MW (and consumption by 1080 Gwh) by 1997. The estimated savings represent 1.7% of projected peak demand ... Thailand is a test site for a comprehensive effort to pursue direct-savings-management in the developing world." Myers, *op. cit.*, 1130.

[52] "Sustainable and Environmentally Sound Development of the Water Resources in Thailand", in UN Economic and Social Commission for Asia and Pacific, *Towards an Environmentally Sound and Sustainable Development of Water Resources in Asia and the Pacific,* Water Resources Series No. 71 (New York: UN, 1992), 201-211.

[53] TERRA has been active in promoting a more open Mekong decision-making process, with both success and frustration. Letter from TERRA to C. Lankester, Executive Agent of the Mekong Secretariat, Feb. 17, 1992. Instead of allowing any NGO to attend, members have insisted on either blanket or, perhaps worse, governmentally appointed NGO representation.

"environment and natural resources of people under oppression, domination and occupation shall be protected."

Seen from this disturbing perspective, the Mekong presents a potential tragedy to us. If the international cooperation which this ancient river has fostered is to have real meaning, it must involve the difficult question of compensating those displaced by Mekong alteration or, less likely, avoiding displacement altogether. More than ever, it is imperative that development, environment, and human rights are viewed as a holistic package, the parts of which are inseparable from each other. From an international perspective, then, Mekong politics present ethical complications to the donor community. Mekong planning implicitly legitimizes the occupation of Tibet and the military dictatorship in Burma; it accepts as inevitable an industrialized region, even though it is quite clear that only Thailand will reap substantial immediate awards; and it involves the international community, most conspicuously through the Asian Development Bank, but also through development assistance programs such as the Canadian involvement in the Low Pa Mong mentioned above, in this scheme.

At the same time, the security function of Mekong cooperation should not be over-emphasized. Plans to "harness" the mighty Mekong have much more to do with facilitating the exploitation of natural resources than they do taming an untrustworthy China. Nonetheless, such perhaps exaggerated strategic factors will continue to play a role in justifying the environmental alteration of the Mekong resource domain. It is somewhat of a Cold War mentality that associates Japanese financial participation with an effort to strengthen a buffer zone in Southeast Asia against a potentially aggressive China, which already has strong links in the region. And if China's troubled international reputation stands only to gain with cooperative initiatives in the region, its mineral and hydropower export sector will gain as well.

However, regarding the strife characteristic of intra-regional relations, sharing the Mekong will be an important instrument for the maintenance of stability. In particular, the conflict between Thailand and Vietnam could swell with dam construction and irrigation diversion projects by the former. The Mekong Secretariat remains a vital communication link in such matters. Likewise, whatever evolves out of the current political developments in Cambodia will find international currency -- in terms of both acceptance and donor support -- by participating in the Secretariat's affairs. Optimistically, the Secretariat could facilitate dialogue between the governments concerned while simultaneously evolving towards a conservation-oriented body that ensures the participation of

NGOs in the policy process, but this may not be possible given its limited autonomy.

Two things seem certain. Firstly, we must take environmental security into account in any analysis of international politics. Doing so challenges, but in no way precludes, state-centric approaches. This is because on the one hand environmental alteration is so relevant at the substate and transnational level, while on the other governments will always be intricately involved and in many cases, such as this one, inter-governmental organizations will play an important role. As this case study suggests, an analytic framework for global environmental security must involve a multitude of levels and dimensions, such as geopolitics (especially, but not solely because of, China's relatively recent involvement), the political economy of development, human rights, state and non-state international organization, ecology, and technology.

Secondly, when it comes to international developmental concerns, it is the renewable resources, such as water, air, trees, and soil, that are most important. Their future treatment will be as vital to both global and regional survival as it will be indicative of international, and human-nature, relations. Thus the importance of the Mekong Commission for Southeast Asia, and, for Northern donor nations, the dilemmas involved in channelling development assistance in its direction.

CHAPTER SEVEN

CONCLUSION: THE PARENTS OF POLITICAL INVENTION

"One of the great challenges facing the human species is to ensure that the human biological characteristic, the aptitude for culture, does not prove to be a prime example of dead-end adaptation."
S. Boyden, 1992[1]

"The concept of Progress as it has developed in the Western tradition displays a glaring and dangerous omission: it conveys no sense of threshold, of irreversibility, of natural limits, or of the sheer scale of human activity running out of control. We seem to be stuck in a combined Promethean, Cartesian, and Faustian attitude of power and control..."
S. George, 1993[2]

1. SECURITY REVISITED

This is, perhaps like all others, an age of paradox. The demands on the international system for agencies and programmes which mitigate the potential damage caused by that very system have never been as great. Contemporary global society challenges the prevalent definition of security in all its aspects and at all levels of analysis. There has been a flourish of international organization, as well as a change in conventional regime politics and a diffusion of the meaning of security, all of which

[1] S. Boyden, *BioHistory: The Interplay Between Human Society and the Biosphere, Past and Present* (Paris: UNESCO, 1992), 41.
[2] Quoted in *New Perspectives Quarterly*, 10:2 (1993), 53-54.

reflects increasing anxiety over continued planetary degradation and environmental degradation. This study has sought to examine some examples of such slow transformation, and to discuss in conceptual terms its apparent significance.

Though the limited nature of this study precludes broad assertions, it could be argued that traditional security issues -- the defense of "homeland", or territorial space, and according to some its expansion -- have been joined in the realm of high politics by questions such as energy planning, resource use, and development options. It may also be argued that government representatives have been joined by other political actors at the stage of agenda setting and, even, the tables of negotiation. This is not a striking phenomenon with regard to industrialists, which as each of our case studies reminds us have always wielded great influence in world affairs and within international organizations; but the proliferation of the participation of other interest groups, such as environmentalists, in world affairs lends some credence to the pluralist perspective on world politics discussed earlier and revisited below.

One might, however, wonder if these are but the images seen on a still shadowy stage, while the play remains the same. Is the fundamental structure of world politics changing? To answer such a question we would need a definitive model of world politics, which we do not have. But we can say that the institutions which have been driving or (conversely) limiting change for the last 500 years, such as state sovereignty, the expansion of capitalism along with state intervention, and the manipulation of natural resources to meet both human needs and excesses, continue to dominate, if not write in its entirety, the script. What is relatively new is an appreciation of the complexity and consequences of decision-making relating to the environment, and this as we have seen has become quickly institutionalized itself, by the international organizations we have studied here and by others. Whether this represents a marked normative shift in how international politics is done is rather more doubtful.

Security as a concept has, this author believes, changed in its prevailing meaning. While it is certainly so that environmental security, or some version of GES, has not become a universally accepted paradigm of thought and action, it has become a permanent fixture in the broader conception of security, for foreign policy and non-governmental actors alike. And though one can always argue that the change this provokes in regimes is but epiphenomenal, it is change nonetheless. However, it would be quite premature to argue, from the environmentalist standpoint, that anything near a conceptual victory has been won. Experts on international organization and law tend to agree. Donald Puchala feels that environmental issues are "fashionable [but] few are genuinely global. Even fewer are amenable to global management." Philippe Sands believes "it will be another twenty years before environmental issues are anything like integrated into the mainstream of economic issues."[3] Despite Stanley Hoffmann's sage parable, a position of humane scepticism should be adopted here.[4] Environmental problems are becoming so noticeable, so persuasive in their urgency and costly in effect, that they are becoming the parents of political invention. In other words, dealing with them, or at least appearing to be dealing with them, has become a necessity, at the domestic and international levels. This is the blunt edge of inspiration, but it is the only one that, in the tumultuous and fragile order of world politics, seems to matter with any consistency.

2. World Politics Revisited

As we began our initial conceptual discussion with a brief overview of three basic strands of international relations theory, it will benefit the present analysis to revisit them.

International cooperation related to environmental problems still centers around the formation, maintenance, and transition of regimes, or levels of international organization. An appreciation of the complexity involved in any such development should dilute the methodological utility of the regime concept, but it remains employable. Functionalists, on the other hand, may be disappointed with the limited autonomy of international organizations, or with the fact that they remain an adequate forum for the articulation of problems but not the enforcers of solutions. Their role as regulators is contingent upon national participation, and subject to challenge and outright defiance. They are quite limited by physical parameters as well. It is hardly sensible to expect the IAEA, for example, to have complete surveillance over all atomic energy plants in effect at all times. Likewise, the IWC 's monitoring effectiveness depends on whaling nations' cooperative participation, granting permission to observers who then board and record the activities of whaling ships (as mentioned in Chapter Five, fabricated Soviet reports in the 1970s demon-

---

[3] Personal correspondence, July and August, 1993, respectively.

[4] "Humane skeptics", Hoffmann writes, "invite nods and sighs, not sound and fury; and sound and fury are good for creative scholarship." "An American Social Science: International Relations", *Daedalas*, 106 (1977), 45.

strate how precarious an enterprise this is). And the Mekong Committee, though it will funnel development funds and is capable of legitimizing or criticizing certain plans for dam construction or river diversion, is powerless to stop local national governments from proceeding with plans. The 1993 Mekong Work Programme, the Secretariat writes, "responds to the national political realities on the one hand and the interests and funding priorities of the donor community on the other."[5] And, of course, the international organizations mentioned above have only limited influence, since no regime is universal. Nuclear development programs remain prime examples of the secrecy attached to such prestigious national investments with military implications. As we saw in Chapter Five, it was an NGO, Greenpeace, which followed and publicized the activities of non-IWC, or "pirate", whalers. And the Mekong River will be polluted by a wide array of industrial and agricultural actors clearly beyond the organizational or jurisdictional capacity of the Committee.

As for the conceptual debate over "hegemonic leadership", we may conclude that leadership played a necessary, but hardly sufficient, role in regime formation. All the regimes and international organizations we have examined were formulated shortly after the Second World War, when the United States was without doubt the preeminent superpower on the globe. Even the Mekong Committee, which didn't receive official status until the late 1950s, began with heavy American involvement, in particular that of the Bureau of Reclamation and Raymond Wheeler, the Chief Engineer of the Army Corps. Funding for the Secretariat had Cold War incentives (the best defence against communism, aid thinking at the time went, was development). Interestingly, American influence does not seem to have been maintained throughout the years, especially after the military involvement in Vietnam, though Thailand and China trade heavily with the United States (and the other Mekong states would probably like to). Japan, which is funding much of the Asian Development Bank plan mentioned earlier, has greater direct influence; and China, as we have seen, will have great influence over the eventual shape of the regime. The heaviest donor to the Mekong Committee itself is the European Community (Union), followed by Sweden and various UN agencies. Hegemony, in its simple non-Gramscian sense of dominant power, seems unlikely. (If there is a hegemonic force guiding Mekong alteration, it is the ideational human-nature ideology discussed below.)

[5] *Annual Report 1992* (Bangkok), 8. The 1993 Programme was substantial, consisting of 116 projects (52% basin-wide, 48% national), with a total budget of $196 (US) million.

The international politics of whaling suggest a similar diffusion of power and influence, involving the highly visible participation of non-governmental actors. The original *International Convention for the Regulation of Whaling* was signed in Washington. In fact it was the American occupying force that granted the Japanese permission to resume whaling, contingent on joining the regime. Despite the understandable inconsistencies related to subsistence whaling mentioned in Chapter Five, it was the American anti-whaling position, coupled with the threat of American economic sanctions, that provided the statecraft necessary to change the normative bias of the IWC. At the same time, the involvement of protest groups and scientists also had immense effect, as well as that of other states-- including relatively powerless, smaller states that, for one reason or another, elected to join the IWC in the 1970s and early 1980s. We should be rather hesitant to adopt the language of confrontational high politics with regard to this issue. Though the United States is prone to threaten the imposition of sanctions (and, in limited cases, has imposed them), it is not prepared to let the issue evolve into a full-blown trade war with the Japanese.

With the nuclear trade/weapons regime, it is also obvious that American governmental representatives, as well as the American nuclear industry, had much to do with the present situation (a regime which attempts to curb proliferation as it promotes the spread of nuclear energy). Not only does the United States remain the main actor in nuclear affairs (as recent negotiations with North Korea clearly demonstrate), it is also the largest contributor to the IAEA. But as we have seen, the nuclear issue became such an internationalized issue because the American nuclear industry lost its monopoly on nuclear power and technology some time ago. In fact there are contemporary fears about "secondary suppliers", including so-called Third World states, in today's troubled nuclear marketplace. Hegemony is simply too strong a word to describe the leadership role played by the United States.

Structural globalists would agree with state-centrists on one thing: international regimes are not going to determine the future of world politics, nor the inequality that characterizes the present global order. There is little in this study to dispute this assertion. Despite its supposed egalitarianism with regard to promoting the spread of nuclear power, the nuclear non-proliferation regime remains a blatant example of how the powerful prefer freezing inequalities to addressing them through perceived sacrifice. The whaling issue may have involved developing nations, but they were brought into the game largely by proponents of non-whaling and, today, by Japanese pro-whaling linkages to development assistance (other issues of the "commons", such as the Law of the Sea negotiations or the Antarctic Treaty, have similar patterns). And the Mekong is being altered largely, one might argue, to serve the interests of the dominant strata within the affected countries, which in turn are involved in export industries which perpetuate global core-periphery linkages by exploiting local labor and resources.

One might temper this conclusion somewhat with reference to the role played by "counter-hegemonic" actors in the cases. In the case of atoms, the peace and environmental movements forge a common link in protesting the spread of nuclear power and weapons, one not without influence even in places as barren of domestic pluralism as the former Soviet Union. The whaling issue, we have argued, served as a catalyst for the articulation of an alternative conception of human-nature relations, a theme we will return to below. And in the case of the Mekong, NGOs such as TERRA push for greater accountability and citizen participation in relevant decision-making processes. These are not dominant forces, nor are they part of dominant structures, but they represent struggle and diversity, and thus challenge the symmetry of the rather neatly modelled core-periphery -- or state-centric -- world. They also force us to broaden our understanding of regime dynamics and, even, the concept of structure itself.

Normative globalism, meanwhile, remains the ideational context of much environmentalist thought. Like any other ideal, it becomes more difficult to keep in focus as the exigencies of daily policy and the constraints of structure come closer. Optimistically, an ontological shift toward biospheric concerns may be occurring in the natural sciences, and even in the social sciences. The latter, however, is far from approaching an influential position. The institutions that drive research and policy are founded largely upon the state, or upon research and development concerns serving parochial interests. It may be possible that a truly functionalist and autonomous international organization can overstep this fact and represent the interests of the globe, but this is at best wishful thinking. This is not to deny the technical capacity of international organization. Each of our studies has shown this exists: that technical expertise can be assigned to investigate scientific questions, such as acceptable levels of radiation, or whale kills, or the sedimentation level of rivers.[6] How accurate these measurements and projections will be will

[6] We should note that this expertise can assist other efforts as well. For example, in 1992 the Mekong Secretariat supplied important data on land availability and suitability to the UNHCR to assist in the massive (370,000) and successful resettlement of Cambodian refugees

depend on the science conducted and the level of commitment, expertise, and so on. But the application of this knowledge will be political in nature: it will represent or support an agenda, and usually that of particular individuals, groups, or institutions, and not the biosphere as a whole. Ecofeminist Vandana Shiva warns us to treat with special suspicion the rhetorical and commercial call for "globalization": "[t]his supposed concern with 'global' issues does not arise out of a concern for all humanity or for all life on earth: instead it reflects a concern to protect the interests of particular local and parochial interests which have become globalized through the scope of their reach."[7] All of this may be a bit unfair. There could be a global stewardship ethic emerging, which accepts such dictums as "responsibility horizons ought to match technological-power horizons."[8] But for the present, a convenient if frustrating compromise must be struck between the reality of parochial politics and the ideal of globalism.

3. GES Revisited

Part of that compromise may be some concept of Global Environmental Security, such as that discussed in this study. GES as outlined in Chapter Three was intended primarily as an analytic device, offering a variety of pertinent aspects for the study of international environmental politics. But, since it does not insist on a jealous adherence to a particular level of analysis, it could be employed from a pure foreign policy perspective as well. How has the introduction of three fairly diverse case studies influenced our original formulation, if at all? What are the implications for the continued validity of the concept, as an analytic framework or as a prescriptive one? A brief discussion will demonstrate that GES as conceived here emerges relatively unscathed from the empirical aspect of this book.

---

who had been encamped near the Cambodian-Thai border. Mekong Committee, *Annual Report* 1992 (Bangkok), 7.

[7] V. Shiva, "Editorial", *The Ecologist*, 22:6 (Nov/Dec 1992), 258.

[8] J. Oberg, "Ecological Ethics: Nature Has Rights, Humans Have Duties", *Bulletin of Peace Proposals*, 21:4 (1990), 403-411, 409.

I. ECOSYSTEM HEALTH

It is obvious that human existence is predicated upon some form of environmental alteration. Deliberate fire was the first use of extrasomatic[9] energy; it may also be construed as the most important advance ever made in technology. But it involved burning trees, either those which had died and were decomposing as part of some natural process, or those which were, again deliberately, pulled down for the purpose of fire. Plato wrote early in the fourth century B.C., referring to the disappearance of forests in ancient Attica: "What now remains compared with what then existed, is like the skeleton of a sickman, all the fat and soft earth having been wasted away, and only the bare framework of the land being left."[10] Plato was watching an early and relatively benign process of clearcutting, a process that would facilitate economic growth, war, roofs for philosophers' heads, and warm baths for senators.

It may be pedagogically useful to teach the history of humanity from the disturbing perspective of a progressively diminishing ecology of the biosphere. The face of the earth has shifted as forests have receded; deserts have advanced in both hemispheres; oil has been spilled and dumped in oceans; and smog-bound cities have multiplied. As sinister are the less visible shifts, in the rapid increase of atmospheric carbon dioxide, in the accelerated destruction of plankton by ultraviolet rays, in the demographic transition in the south. Threats to environmental security, as the WCED pointed out, are linked to each other as well as to economic and political factors. Problematically, such "links mean that several different problems must be tackled simultaneously." However, and in a more positive light, success "in one area, such as forest protection, can improve chances of success in another area, such as soil conservation."[11] This is easily said. Realizing that the problems of nuclear proliferation and the risk of radioactivity, especially in the Third World, are part of a larger problem of North-South inequity does not make solving any of them automatic. Realizing whales were close to extinction due to overhunting does not imply that the related problem of whale habitat destruction will be met by international organization. And an awareness of the negative effects of large dam construction will not necessarily slow the quest for rapid industrialization in the Mekong area. What we can hope

[9] As opposed to somatic energy, which is consumed in food directly by an organism.

[10] Boyden, *op. cit.*, 127. Boyden also quotes Homer, who likened the noise of battle to the "din of wood cutters in the glades of a mountain."

[11] WCED, *Our Common Future* (New York: Oxford University Press, 1987), 37-38.

for is some cognitive progression as a result of making these links clearer, and this will be a political phenomenon.

II. RISK AVERSION

In all of our case studies, we have seen how international organizations legitimize certain forms of economic behavior (nuclear trade, whaling, and environmental alteration) which may have an adverse impact upon ecosystem health. We have also seen that these organizations can assume positions with more or less pro-ecosystem biases, depending largely on who seems to be influencing them at the time. This would at least imply, to adopt the rather technical terminology of the cognitive theorist Ernst Haas, that these organizations have engaged in "adaptation", if they are not necessarily "learning."[12] The IAEA does spend more time and money on environmental issues now than ever in its history; the IWC has for some time now taken conservation seriously as a mandate, indeed too seriously for some; and the Mekong Commission makes sure its publications refer to the environmental impact assessment processes in which it participates.

Contemplating future risks imposes greater demands on decision-making. There is a certain innocent audacity involved in all this. A scientist working for Atomic Energy of Canada Limited wrote in 1953 that the storage of radioactive waste "must be impervious to winds and water erosion for centuries to come", and that storage in tanks was possible, though "[i]t would be necessary to monitor the environment for *centuries* to be sure leakage was not occurring."[13] For the nuclear complex, born of technological fixation, whether the political and economic circumstances would facilitate such monitoring for centuries was not an important issue; that future generations would have to bear the costs of this disposal was not a serious consideration. The question takes an interesting twist today, since it has become increasingly difficult to convince those most astute of decision-makers -- insurance executives -- to take risks on nuclear power programmes, especially in Eastern Europe. Unfortunately, this is deterring the badly needed modernization of these plants. As Martin List

---

[12] See E. Haas, *When Knowledge is Power: Three Models of Change in International Organization* (Berkeley: University of California Press, 1990).

[13] A.M. Aikin, "Disposal or Uses of Large Quantities of Fission Products", in AECL *Report on Atomic Power Symposium,* held at Chalk River, Ontario, 9-11 Sept. 1953 (CRR-548-A). Italics have been added.

writes, Western firms "shy away from the associated risk of incurring liability for consequent damage in the event of an incident in Eastern nuclear power plants retrofitted with Western technology."[14]

The complexity of the case studies examined here points to the impossibility of identifying precise locations of decision-making. Rather, political behavior is the result of an accumulation of decisions made under certain conditions. This does not imply that international organization, and environmental politics, do not induce certain patterns of decision-making. For example, there is a marked tendency to react to environment problems if and when they become undeniable and raise sufficient public concern in liberal democracies or, even, less tolerant political climates. There is also a strong need to legitimize risk taking through the discourse of international diplomacy. This will probably continue in the near future; a cynical interpretation of UNCED might consider the entire process a public relations campaign designed to ensure things carry on much as they have been. A closer look at environmental regimes reveals real changes, perhaps even significant ones, with the advent of environmental concerns; but we should be always be aware of the promotional dimension of regime activity.

As well, we should keep in mind what may well be one of the few truisms in politics, the fact that the share of risk is usually disproportionate. In other words, decision-makers themselves don't always live the consequences of the risks they take. It can be argued that we all lose in a nuclearized world, or that we can all gain from nuclear power, but we do not all live near nuclear plants, and none of us will be present in the future when the atomic waste question will have to be definitively answered. In the case of whaling, it was the conscious decision to act on behalf of whales -- to internalize the risk to their survival -- that generated so much anti-whaling sentiment. But it would be chimerical to imply this was some universal development: to some, the biggest risk involved with relation to whaling is ensuring a future supply of fur-farm food. And the Mekong is an excellent example of how ecological risk will be felt most acutely, regardless of the exact nature of the impact of environmental alteration, by the poor, the riparian people disconnected from the emerging metropoles and dependent upon subsistence agriculture and fishing. Of course, the Mekong Committee, and others, involve some representa-

---

[14] List continues: "[t]he question of whether and under which conditions a public Western coverage of the liability risk could or should be ensured is currently being negotiated in the G7 Working Group entrusted with this task." *Op. cit.*, 352.

tion of such groups in their policy planning. But risk, like human rights themselves, is a relative thing.

Finally, it is important to recall the importance of the market as a structural decision-inducing element. In all the cases examined here, we see the importance of the market as a global form of social organization. This relates to the contemporary fears about the "buyer's market" in nuclear technology, as well as its initial spread as domestic consumption fell in key producing states; to the foolish slaughter of whale species such as the Blue Whale, and current calls for renewing sustainable harvests of minke whales; and to the rapid growth of export-oriented industry and agriculture in Thailand and China, as well as the increasing market for energy amongst Mekong area states. Indeed, the Chinese Government's intention to dynamite the Mekong rapids to facilitate commercial navigation downriver is driven by world demand for phosphorous found in Yunan province. The world economy, with its many complex and overlapping layers of production, distribution, and consumption, has played a key role in all these developments.

III. Human Rights

The human rights dimension of GES is the most manipulable. That is, it is possible to argue from most sides of an environmental risk decision-making dilemma that there is a human right involved. Those opposed to nuclear power argue that people have the right to an environment free from the risks associated with it, and with the possible spread of nuclear weapons. Others may argue that Third World nations have the right to energy sector development, as well as the right to choose for themselves what technology will be employed. Similarly, Indian or Pakistan officials might argue their citizens have the right to be protected from nuclear attack, and that this necessitates the establishment of a nuclear deterrent. Anti-whaling activists have argued that humans have a right to biodiversity, and to the experience of sharing the globe with other mammals. Others may argue that whalers have a right to employment; or that whaling nations have the right to determine their own destiny (the older argument that whaling would feed the world's poor, and that denying whalers their trade was thus perpetuating chronic hunger and disease, has been mercifully dropped from such polemics). And in the case of projects which will displace riparian residents of the Mekong River basin, it is quite straightforward that they will be losing their human rights to live peacefully where they have, often, been located for centuries. Downstream states and peoples may argue they have the right not to be affected by upstream developments, and international legal convention partially supports this. Industrialists will argue that the countries involved have the right to industrialize, that in the end some small sacrifice must be made in order to satisfy this right for the majority. The complications are seemingly endless.

As well, each of the studies examined above has a temporal human rights component: the right of future generations to live without fear of radioactive waste, without a simplified marine ecosystem, and without the permanent loss of an ecosystem such as the Mekong River Basin. It can be argued that this is the most sensible version of the rights question, though this may sound far more technical than the discourse on human rights usually prefers. Finally, we should not forget that, from the viewpoint of either globalist humanism or state-centric pragmatism, inequality itself may be seen as a global security problem. Craig Murphy feels that "political and economic inequalities cause the greatest immediate 'security' problems, in particular, the trend over the last 20 years of the wealthy getting wealthier."[15] This trend has been partly facilitated by the continued exploitation of the environment.

Not only is the human rights framework troubled by the secular/temporal split, but it must answer to charges of anthropocentrism, a theme we will return to below when we discuss human-nature ideology. Overall, and though it remains useful to discuss human rights as a component of environmental security, we are reminded of the subjective nature of any social concept. Later in this Chapter we will discuss in greater detail the normative context of GES.

IV. SOVEREIGNTY IN WORLD POLITICS

Sovereignty may have changed in character, reflecting the growing interdependence of economies or even the continuation of global imperialism, but it remains a central institutional tenet in world politics. The decision-making processes in each of the studies used here rely to a great extent on the participation of intergovernmental actors which are aware of this, and usually have explicit instructions from national authority centres which predetermine their behavior. Long before sovereignty falls as an

---

[15] Personal correspondence, Craig Murphy, Department of Political Science, Wellesley College, Mass., June 1993.

institution, environmental problems will reach a point where they provoke increased international conflict. This state of affairs is frightening because it is not an abstract projection, but is actually occurring in limited cases, such as water disputes. The debate over whaling has served to exacerbate differences between nation-states, in particular between the newly non-whaling United States and Japan, who are also of course the two major trading rivals in the global political economy. And the issue of nuclear inspections is by definition a sovereignty issue.

As resources dwindle, sovereignty assumes even greater importance. In cases where the commons are at stake, this will necessarily involve international cooperation. However, the unilateral approach may well gain credence as the most pragmatic with regard to resources within reach of national governments. Canada, for example, has recently declared a willingness to block non-Northwest Atlantic Fisheries Organization (NAFO) members from sending trawlers into the Grand Banks off Newfoundland. The dispute is a rather nasty one, since these foreign vessels are largely suspected to be NAFO member-owned, reflagged in order to escape NAFO provisions (the entire EC has joined NAFO). The "nose and tail" of the banks are beyond the 200-mile nautical-mile limit; Canada would in effect be patrolling the commons in order to enforce the dictates of the NAFO. It is too early, at this stage in the new Administration elected in Ottawa (early 1994), to say with any certainty that the Canadian Government will enforce this foreign policy position, though it has apprehended a Canadian-owned trawler fishing off the Banks with an expired flag of convenience. Multilateral solutions will always be sought to avoid overt conflict and build confidence in the future. But it is clear that unilateral options are open for consideration.[16]

Aside from reconceptualizing the value of sovereign territory to include its environmental worth or sustainability, a process which seems well under way (though not by any means uniformly), governments remain reluctant to sacrifice what are perceived as sovereign prerogatives.

---

[16] P. Koring, "Canada to Block Fish 'Pirates'", *The Globe and Mail*, Jan. 12, 1994, A1 and A2. Canadian Fisheries Minister Brian Tobin, who made this announcement during a meeting with EC Fisheries Commissioner Vannis Paleokrassas, told reporters: "It's an act of conservation; nobody could call it an act of war. We're out to declare enough is enough when it comes to the desecration of cod stocks by nations that operate outside of any civilized norms"[!]. *Ibid.* On this issue I am grateful to Robert Applebaum, Director General of the International Directorate of Fisheries and Oceans, Government of Canada, for a discussion in January of 1994; and to the optimistic Dr. Schepel, Executive Secretary of NAFO, for a related discussion. (Subsequent to the original drafting of this bok. Canadian foreign policy on this issue has become increasingly assertive with regard to Spanish turbot fishing vessels.)

The North-South divide, so clearly illustrated by GES considerations like global warming and soil erosion, does not preclude the infinite levels of intergovernmental interaction, the seemingly endless plethora of diplomatic protocol, the painfully slow processes of international law, the grandstanding afforded by governmental summitry, or the other multitudes of conventions and traditions that go into the institution of sovereignty today. This makes for slow, indeed reluctant, transformations at the global level, even with regard to seemingly small and practical issues, such as the protection of specific species, or the environmental alteration of geographically defined ecosystems.

The fear that environmental problems, and resource problems in particular, will accentuate the least desirable aspects of sovereignty is a reasonable one. Douglas Ross, for example, fears the "siege mentality/garrison state perspective" is gaining ground.[17] The cure for this affliction remains cooperation, and this will continue to be focused at the level and in the activity of international organization. This implies a willingness to subvert the theology of sovereignty to various degrees. Once this admittedly liberal-institutionalist premise is accepted, the real question concerns the type -- the extent and influence -- of the machinery and architecture developed.

The most obvious prescription is for an institution capable of meeting the global nature of the challenge; one which is able to reorient global order towards concepts like sustainable development. This may vary from the present UN system itself, which is slowly turning to reflect sustainability as a guiding principle, even if it is not acquiring the means to implement it, to new, universal creations. Such grand thinking is highly conjectural at this stage. It is for example difficult to conceive the explicit institutionalization of values held by "environmentalists" (if there are a coherent set of such values) in some sort of Institutional Environmental Organization, such as that recently proposed by D. Esty.[18] It was straining enough to arrive at the Rio Declaration, a watered-down version of the originally planned Earth Charter. It is arguable that it is also dangerous to think in terms of single-stroke solutions based on highly central-

---

[17] Personal correspondence, July 1993. Ross teaches political science at Simon Fraser University, B.C.

[18] D. Esty "GATTing the Greens, Not Just Greening the GATT", *Foreign Affairs* 72:5 (1993), 32-37. Esty helped negotiate the environmental provisions of NAFTA. IEO would "protect environmental values the way the GATT guards free trade principles" (34). For an earlier call, see G. Kennan, "To Prevent a World Wasteland: A Proposal", *Foreign Affairs*, 48 (1970), 401-413.

ized political authority, exaggerating their potential for order and underestimating the importance of sub-state, localized movements.

It may be more sensible to pursue regional agreements, such as the Ramscar Convention on Wetlands of International Importance, the Bern Convention on the Conservation of European Wildlife and Habitats, and the Geneva Protocol Concerning Mediterranean Specially Protected Areas;[19] or to sophisticate more permanent regional arrangements, such as the Mekong Development Committee. It may be possible in the long run to pursue concise regimes related to a certain type of ecosystem, such as a proposed International Marine Conservation Network.[20] This would not, of course, eliminate the problems posed by the territorial separation of sovereign states, including the thorny issue of verification. (Indeed, the arms control literature will probably be the source of terminological inspiration in future accounts of environmental diplomacy.)[21] If such a Network obtained some sort of autonomous decision-making capacity, and it decided (quite rightly) that France-Japan plutonium shipments posed a grave risk to environmental security, would the world community be prepared to impose appropriate sanctions against those nations?

There are also practical limits relative to the size of the social problems caused by ecological crises. Regardless of the validity of worst-case scenarios concerning the effects of global warming, environmental decay -- in particular, the diminution of resources such as soil, trees, and fish -- will undoubtedly give rise to unprecedented levels of population displacement and mass migration, much of it across borders. The UNHCR has reacted by establishing a single analytic post at its headquar-

---

[19] Respectively: UNTS 14583 (1971); 21159 (1979); and 24079 (1982). See also A. Westing, "Biodiversity and the Challenge of National Borders", *Environmental Conservation* 20:1 (1993), 5-6. Such arrangements remind one of the calls for geographically-based disarmament; for an early version, see J. H. Marshal-Cornwall, *Geographic Disarmament: A Study of Regional Demilitarization* (London: Oxford, 1935).

[20] See E. Norse, *Global Marine Biological Diversity: A Strategy For Building Conservation Into Decision Making* (Washington: Centre for Marine Conservation, 1993). Reviewer Don McAllister calls this book a "major seamark" in the literature. ("For those who haven't guessed, seamark is the marine equivalent of landmark - a plethora of words like landmark expresses the terrestrial bias of the human ethos, a bias that must be overcome if we are to fully appreciate, protect and wisely use marine biodiversity."[!]) Review is in *Sea Wind: Bulletin of Ocean Voice International*, 7:3 (1993), 34-35.

[21] For example the U.S. Accounting Office has released *The Verification of International Conventions on Protection of the Environment and Common Resources: A Comparative Analysis of the Instruments and the Procedures for International Verification with the Examples of 13 Conventions* (Washington, 1992).

ters in Geneva.[22] Clearly, the UNHCR is most understandably preoccupied with other acute refugee crises related to military warfare. The real solutions to the problems causing by mass population movements (if there are any at this stage) will require incredible sums of assistance, and it is highly unlikely that increased funding is forthcoming from the recession-weary North.[23] Human adaptability, then, will face its greatest challenges in the future.

4. Human-Nature Ideology

Our case studies have a very common thread: attitudes towards the environment which perceive it as a resource intended for human consumption have prevailed in the post-World War Two era. The development of atomic weaponry and power exemplifies this approach, reliant on what has been popularly termed "big science." Perhaps the most poignant example of this was discussed in Chapter Four, in relation to the Manhattan Project. The understandably hurried development of the bomb and its essential military mission "left a political and organisational legacy that meant decision-making over atomic energy was for the next three decades (1945-75) dominated by three components: secrecy, centralization and technocracy. These led respectively to lack of political or popular participation; to institutional self protection; and to a lack of properly debated policy."[24] Over 150,000 personnel were involved in the Manhatten project, and only a handful understood the total impact of their collective effort. Evidence exists which suggests that the U.S. Military was more concerned, initially, with the radioactive potential of nuclear waste (fission products) than with the destructive potential of an atomic weapon. Indeed it is even claimed -- and the claim is supported by declassified USDOE documents released in 1986 -- that the citizens of Hanford were used as guinea pigs for radiological experiments.

---

[22] This is the Coordinator of Environmental Affairs, currently held by Kazuo Watanabe, with whom I have been in contact.

[23] Personal correspondence with Qazi Shaukat Fareed, Director, Department of Humanitarian Affairs, United Nations, New York, July 1993. Fareed insists nonetheless that enlightened self-interest remains an incentive. As Fridtjof Nansen, who pioneered refugee assistance programmes, is supposed to have said, "love of man is a practical policy."

[24] A. Blowers, D. Lowry, and B. Solomon, *The International Politics of Nuclear Waste* (London: MacMillan, 1991), 35. See also R. Rhodes, *The Making of the Atomic Bomb* (Harmondsworth: Penguin, 1988).

This is no blanket condemnation of science, or especially of those who are called scientists, many of whom have been active in promoting the idea that science can't do everything, and, in Oppenheimer and Einstein's tradition, have frequently spoken publicly against military projects often involving both nuclear materials and outlandish schemes. Just recently a group of American scientists expressed their dismay over plans to put additional Russian nuclear reactors into orbit to facilitate research on SDI; the scientists point to the 1978 crash of a small reactor in northern Canada, and claim that reactors will "disrupt the operation of other science missions."[25]

Perhaps the nexus of human-nature relations is the way the environment is viewed as a resource. All the case studies deal, directly or indirectly, with the question of resources, and even more specifically, energy (we should not forget that one of the earliest sources of industrial energy, with the discovery and subsequent extraction of sperm oil, was the whale). The institutions that formed around these resources were designed primarily to facilitate their exploitation by interested parties, be they the nuclear energy sector, the whaling industry, or the firms that have invested in agricultural production in the Mekong delta and in industrial production in the riparian states. All three organizations, the IAEA, the IWC, and the Mekong Committee, were formed following World War Two, which signalled a new beginning in the history of international organizations, one in which technical promise cast an optimistic shadow over the potential risks associated with environmental alteration.

Attitudes toward whales - or towards killing them - provide an interesting example of such perceptual shifts of attitudes towards nature, though this is admittedly a comparatively romantic and non-universal instance. Environmentalists might consider it a happy coincidence that power and ethics have fused to create an anti-whaling American foreign policy. Others, including of course the Japanese and Norwegians, view this as a case of the imperialist imposition of (relatively recently acquired) Western values. There is resentment attached to this view, as observers to the IWC have indicated to the present author. At the same time, the state-centric level of analysis, which tends to characterize citizen attitudes as monolithic, distorts the fact that within non-whaling nations there are

[25] The Soviets used reactors -- some 30 of them -- in space during the Cold War to power spy satellites which monitored Western naval operations. The newly proposed reactors would be used to "test nuclear energy as a generator of electricity for operating long-duration propulsion systems and high powered sensors considered essential in surveillance and missile tracking." "Scientists Alarmed Over U.S. Plans of Satellite Carrying Reactor", *The New York Times*, Dec. 6, 1992, 36.

pro-whaling elements, and vice-versa. A whale-watching industry is gaining momentum in Japan, and it will be interesting to see if a pro-whaling foreign policy will survive into the next century. Judging by current Japanese policy, it will. But an undeniable shift in values has taken place. This comes to occasional bright light with small episodes of conscious eradication, such as the American air force firing upon schools of killer whales, ostensibly to help Icelanders get rid of the Orcan menace in the 1950s and for simple target practice in the 1960s.[26] This occurred when whaling was still a legitimate economic activity in most areas of the world; as a normative transition took place, reflected by initiatives at the Stockholm Conference and eventually by an expanded and pro-abolitionist IWC, such peace-time military behavior would be unthinkable.

The nuclear question is much more complex, but again there may be a discernible trend towards refusing to accept the potentially genocidal framework of nuclear weapons-based defense. This framework involved organizations like the IAEA, which somewhat mitigated the anxiety of a nuclearized world. Thus, the paradigm and the institutions became blurred in the bipolar world.[27] Beyond the ethical question of psychological terror, the question of nuclear power involves other normative factors. Environmentalists are concerned with the political implications of such a concentrated, or centralized, energy source. Ecologists are concerned with the long-term consequences of radioactive waste disposal, referring again to the temporal dimension of human rights, and the various risks associated with the nuclear fuel cycle, including the long-range transportation of plutonium. The bottom line of nuclear anxiety, however, has been the less analytic question: are we pushing nature too far?

The Mekong Committee, meanwhile, facilitates the process of industrialization, or what some prefer to call modernization, which is often at odds with the human-nature conception of many of the areas Buddhist inhabitants. Rather than a source of cultural continuity, the Mekong has become a source for electric power, irrigation, and navigation schemes. Resistence to the dictates of the neo-liberal globalization agenda may or may not be directly related to ecological concerns, but they certainly form part of the counter-hegemonic argument expressed by NGOs. The

---

[26] This practice has been widely reported; see for example D. Day, *The Whale War* (Vancouver: Douglas and McIntyre, 1987), 5-6.

[27] Students of genocide have tended to agree: Robert Lifton and Eric Markusen, for example, discuss how a certain "genocidal mentality can become bound up with the institutional arrangements necessary for the genocidal act." *The Genocidal Mentality: Nazi Holocaust and Nuclear Threat* (New York: Basic, 1990), 3.

Committee, born of the golden age of engineering, has faithfully promoted the ability of modern infrastructure projects to bring material wealth and create fertile ground for regional cooperation. Importantly, however, this is complimented by initiatives such as attempts to save Lower Mekong Basin wetlands, and to monitor water quality.

It may be appropriate to conclude this study with reference to a more modest species than *Homo sapien*. S. Boyden writes of what he terms the "flaming moth principle":

> at night-time moths often fly directly into exposed flames, thereby incinerating themselves. This behaviour, which is apparently an expression of an innate behaviourial characteristic of these insects, is clearly not, under the [new] circumstances, of biological advantage to the moth. Nevertheless, the instinct which underlies this behaviour is the product of natural selection, and consequently can be assumed to be of selective advantage in the environment in which moths normally live and in which naked flames are a rare occurrence. In the changed or evodeviant situation, however, the instinct represents a distinct survival disadvantage to individual moths, and to moth populations.[28]

All of the case studies examined here demonstrate the painful learning -- or, perhaps as accurately, delearning -- process that can be partly facilitated by international organization: the learning of limits. It is, by all admissions, a slow process, and hardly a sure one. But there is no need to adhere to Boyden's scenario for self-destruction; we can avoid the fire.

5. A Future Research Agenda

As a loosely defined academic discipline, security studies must recognize the fundamental importance of environmental considerations as they affect all three of the main perspectives on world politics we have discussed. It is no longer possible to logically think of security in purely territorial or military terms. This gives students of the discipline challenging horizons of intellectual and empirical inquiry.

---

[28] *Op. cit.*, 64.

It is hoped that some conception of GES can become a permanent fixture in the analysis of international security: that the threats to ecosystems can be identified; centers of risk-related decision-making analyzed; human rights questions addressed; and the complications arising out of the existence of territorial sovereign units discussed. There is, unfortunately, no shortage of possible topics. The international community has been swamped with the breadth of contemporary environmental problems and, despite attempts to wrap them up in an attractive package such as UNCED or the later Conferences designed in the UNCED process, the problems remain scattered and often unknown to each other. It is the academic's role to bring them together and illustrate various linkages not only between ecological variables but between realms of human behavior as well. That of international organization, from the perspective of global security, remains the most immediately germane.

SELECT BIBLIOGRAPHIES

Note: sources have been listed under the heading of the chapter in which they first appear, and only there in order to avoid repetition. Research for this study was conducted primarily at Douglas Library, Queen's University; Robarts Library, University of Toronto; and the Canadian Institute for International Affairs, Toronto.

CHAPTER ONE: FORGING THE PROSCENIUM

Caldwell, L. *International Environmental Policy: Emergence and Dimensions,* Second Ed. Durham, North Carolina: Duke University Press, 1990.

Engel, J.R. and **J.G.** *Ethics of Environment and Development: Global Challenge, International Response.* Tucson: University of Arizona Press, 1990.

George, S. "Managing the Global House: Redefining Economics." In C. Hartman and **P. Vilanova,** ed., *Paradigms Lost: The Post Cold War Era* (London: Pluto, 1992), 111-129.

Gilpin, R. *War and Change in World Politics.* Cambridge: Cambridge University Press, 1981.

Hampson, F. O. "Peace, Security, and New Forms of International Governance." In C. **Mungall** and **D. McLaren,** ed., *Planet Under Stress: The Challenge of Global Change* (Toronto: Oxford, 1990), 301-317.

Halpern, S. "The UNCED: Process and Documentation." *ACUNS Reports and Papers* (The Academic Council on the United Nations System), 1993.

Hossie, L. "Slipping Environment Onto the Agenda." *The Globe and Mail,* May 26, 1993, A10.

Kegley, C. and **E. Wittkopf.** *World Politics: Trend and Transformation, Fourth Ed.* New York: St. Martin, 1993.

——— *The Global Agenda: Issues and Perspectives.* Englewood Cliffs, NJ: Prentice Hall, 1988.

Meadows, D., et. al. *Beyond the Limits: Confronting Global Collapse, Envisioning a Sustainable Future.* Post Mills, Vermont: Chelsea Green, 1992.

——— *The Limits to Growth.* New York: Times Mirror, 1972.

MacNeill, J., P. Winsemius, and **T. Yakushiji**. *Beyond Interdependence: The Meshing of the World's Economy and the Earth's Ecology.* New York: Oxford, 1991.

Orr, D., and **M. Soroos**, ed. *The Global Predicament: Ecological Perspectives on World Order.* Northern Carolina University Press, 1979.

Pirages, D. *Global Ecopolitics: The New Context for International Relations.* Massachusetts: Duxbury, 1978.

Plant, G. "Institutional and Legal Responses to Global Environmental Change." In I. **Rowlands** and **M. Greene**, eds., *Global Environmental Change and International Relations* (London: Macmillan and Millennium, 1992), 122-144.

Porter, G., and **J. Welsh Brown**. *Global Environmental Politics.* Boulder: Westview Press, 1991.

Rowlands, I. "Environmental Issues in World Politics." In J. Boylis and N. Rengger, ed., *Dilemmas of World Politics: International Issues in a Changing World* (Oxford: Clarendon, 1992), 287-309.

Sands, Philippe. "Enforcing Environmental Security: The Challenges of Compliance with International Obligations." *Journal of International Affairs* 46:2 (1993), 367-390.

Schell, J. *The Fate of the Earth.* New York: Knopf, 1982.

Schriver, N. "International Organization for Environmental Security." *Bulletin of Peace Proposals*, 20:2 (1989), 113-122.

Tolba, M.K., ed. *Evolving Environmental Perceptions: From Stockholm to Nairobi.* Toronto: Butterworths, 1988.

UN. *Earth Summit, Agenda 21: The UN Programme of Action From Rio.* Final Texts of Agreements Negotiated by Governments at the United Nations Conference on Environment and Development, 3-14 June 1992, Rio de Janeiro, Brazil. New York: UN, 1993.

UNEP. *Selected Multilateral Treaties in the Field of the Environment.* Ed. by I. Rummel-Balska and S. Osofo. Cambridge: Grotius, 1991.

Vernadsky, V.I. "The Biosphere and the Noosphere." *American Scientist*, 33 (January, 1945), 4.

WCED. *Our Common Future.* Oxford: Oxford University Press, 1987.

CHAPTER TWO: THERE'S MUCH IN A WORD

Addo, H. *Transforming the World Economy? Nine Critical Essays on the New International Order.* Toronto: Hodder and Stoughton, 1984.

Amin, S. "A Propos the 'Green' Movements." Trans. by P. Burgess and C. Blower. In **H. Addo,** et. al., *Development and Social Transformation: Reflections on the Global Problematique* (London: Hodder and Stoughton with the UN University, 1985), 270-281.

Ashley, R. *The Political Economy of War and Peace: The Sino-Soviet-American Triangle and the Modern Security Problematique.* New York: Nichols, 1980.

Benedick, R. E. *Ozone Diplomacy: New Directions in Safeguarding the Planet.* Cambridge: Harvard, 1991.

Bennett, O., ed. *Greenwar: Environment and Conflict.* London: Panos Institute, 1991.

Booth, Ken. *Strategy and Ethnocentrism.* New York: Holmes and Meier, 1979.

——— *New Thinking About Strategy and International Security.* London: HarperCollins, 1991.

Brewer, A. *Marxist Thought on Imperialism: A Critical Survey.* Boston: Routledge and Kegan Paul, 1980.

Brock, L. "Peace Through Parks: The Environment on the Peace Research Agenda." *Journal of Peace Research,* 28:4 (1991), 407-423.

Brock-Utne, B. "Formal Education as a Force in Shaping Cultural Norms Relating to War and the Environment." In A. Westing, ed., *Cultural Norms, War and the Environment* (Oxford: Oxford University Press, 1988), 83-100.

Brown, C. *International Relations Theory: New Normative Approaches.* New York: Harvester-Wheatsheaf, 1992.

Brown, L. "Redefining National Security." Washington: Worldwatch Institute Paper No. 14, Washington, 1977.

Brown, N. "Ecology and World Security." *The World Today,* 48:3 (1992), 51-54.

Brown, S. "Feminism, International Theory, and the International Relations of Gender Inequality." *Millennium,* 17 (1988).

Buzan, B. *People, States and Fear: An Agenda For International Security Studies in the Post-Cold War Era.* London: Harvester Wheatsheaf, 1991.

——— "New Patterns of Global Security in the Twenty-First Century." *International Affairs,* 67:3 (1991), 431-451.

Byers, B. "Ecoregions, State Sovereignty and Conflict?" *Bulletin of Peace Proposals*, 22:1 (1991), 65-76.

Campbell, D. *Writing Security: United States Foreign Policy and The Politics of Identity*. Minneapolis: University of Minnesota Press, 1992.

Choucri, Nazli. "The Politics of Global Environmental Change: A Conceptual Framework." *UN Advanced Technology Assessment System: Environmentally Sound Technology for Sustainable Development* (Issue 7, Spring 1992), 230-234.

Colinvaux, P. *Fates of Nations: A Biological Theory of History*. New York: Simon and Schuster, 1980.

Cox, R. *Production, Power, and World Order*. New York: Columbia University Press, 1987.

Crosby, A. *Ecological Imperialism: The Biological Expansion of Europe, 900-1900*. Cambridge: Cambridge University Press, 1986.

Dalby, S. "Security, Modernity, Ecology: the Dilemma of a Post-Cold War Security Discourse." *Alternatives*, 17, 1992.

Deudney, D. "The Mirage of Eco-War: The Weak Relationship Among Global Environmental Change, National Security and Interstate Violence." In I. Rowlands and M. Greene, *Global Environmental Change and International Relations* (London: Macmillan, 1992), 169-191.

——— "Environment and Security: Muddled Thinking." *Bulletin of the Atomic Scientists*, April 1991, 22-28.

——— "The Case Against Linking Environmental Degradation and National Security." *Millennium* 19:3 (1990), 461-476.

Diamond, I. and **G. Orenstein**, eds. *Reweaving the World: The Emergence of Ecofeminism*. San Francisco: Sierra Club, 1990.

Eberstadt, N. "Population Change and National Security." *Foreign Affairs*, 70:3 (1991), 115-131.

Engels, F. *Dialectics of Nature*. Fifth Printing. Moscow: Progress, 1972.

Falk, R. *The Promise of World Order: Essays in Normative International Relations*. Philadelphia: Temple University Press, 1987.

——— *This Endangered Planet: Prospects and Proposals For Human Survival*. New York: Random House, 1971.

——— "The Capacity of International Law to Respond to the Environmental Challenge." Address to the Conference of the Canadian Council on International Law: Preserving the Global Environment; Ottawa, Oct. 19-21, 1989.

Finger, M. "New Horizons for Peace Research: The Global Environment." In J. Kakonen, ed., *Perspectives on Environmental Conflict and International Politics* (London: Pinter, 1992), 5-30.

Fortmann, M., and W. George. "Bridge Over Troubled Waters:Linking Strategic Studies and International Relations." In D. Haglund and M. Hawes, ed., *World Politics: Power, Interdependence and Dependence* (Toronto: Harcourt Brace Jovanovich, 1990), 456-481.

French, H. "After the Earth Summit: The Future of Environmental Governance." *Worldwatch Paper* 107 (1992).

Galtung, Johan. *Environment, Development, and Military Activity:Towards Alternative Security Doctrines.* Oslo: Universitetsforlaget, 1982.

Gleick, Peter. "Environment and Security: The Clear Connections." *Bulletin of the Atomic Scientists,* 47:3 (1991), 16-21.

Gonick, L, S., and E. Weisband, ed. *Teaching World Politics: Contending Pedagogies for a New World Order.* Boulder: Westview, 1992.

Gray, C. "Through a Missile Tube Darkly: 'New Thinking' About Nuclear Strategy." *Political Studies,* XLI:4 (1993), 661-671.

Grieco, J. "Anarchy and the Limits of Cooperation: A Realist Critique of the Newest Liberal Institutionalism." *International Organization,* 42 (1988), 485-507.

Grinevald, Jacques. "The Greening of Europe." *Bulletin of Peace Proposals,* 22:1 (1991), 41-47.

Groom, A., and P. Taylor, eds. *Functionalism: Theory and Practice in International Relations.* London: University of London, 1975.

Gurtov, M. *Global Politics In the Human Interest.* Boulder: Rienner, 1988.

Haas, E. "Words Can Hurt You; Or, Who Said What to Whom About Regimes." In S. Krasner, *International Regimes* ((Ithaca: Cornell University Press, 1983), 23-59.

——— *Beyond the Nation-State.* Stanford: Stanford University Press, 1964.

Haas, P. "Introduction: Epistemic Communities and International Policy Coordination." *International Organization,* 46:1 (1992), 1-35.

——— "Do Regimes Matter? Epistemic Communities and Mediterranean Pollution Control", *International Organization,* 43:3 (1989), 377-403.

Haglund, D. "Changing Concepts and Trends in International Security." Paper presented to the Conference on Canada's International Security Policy, York University, Toronto, October 1992.

Halliday, F. "International Relations: Is There a New Agenda?" *Millennium,* 20:1 (1991), 57-72.

Hardin, G. "Living on a Lifeboat." *BioScience,* 24:10 (1974), 561-568.

Harvey, D. *The Condition of Post-Modernity: An Enquiry Into the Origins of Cultural Change.* Cambridge, MA: Blackwell, 1989.

Held, V. "Gender as an Influence on Cultural Norms Relating to War and the Environment." In A. Westing, ed., *Cultural Norms, War and the Environment* (Oxford: Oxford University Press), 44-52.

Helman, U. "Environment and the National Interest: An Analytical Survey." *Washington Quarterly*, 13:4 (1990), 193-206.

Hillal Dessouki, A. "The Two Spheres of Security." *The Washington Quarterly*, 16:4 (1993), 109-117.

Holst, J. "Security and the Environment: A Preliminary Exploration." *Bulletin of Peace Proposals*, 20:2 (1989), 123-128.

Homer-Dixon, T. "On the Threshold: Environmental Changes and Causes of Acute Conflict." *International Security*, 16:2 (1991), 76-116.

——— "Environmental Change and Environmental Security." *Behind the Headlines* (CIIA), 48:3 (1991).

Huelshoff, M., and **T. Pfeiffer.** "Environmental Policy in the EC: Neofunctionalist Sovereignty Transfer or Neo-realist Gate-keeping?" *International Journal*, XLVII:1 (1992), 136-158.

Hufner, K. and **J. Naumann**. "Are the Moral and Value Foundations of Multilateralism Changing?" *International Political Science Review*, 11:3 (1990), 323-334.

Johansen, R. "Toward Post-Nuclear Global Security: An Overview." In B. Weston, ed., *Alternative Security: Living Without Nuclear Deterrence* (Boulder: Westview, 1990), 220-260.

——— "A Policy Framework for World Security." In M. Klare and D. Thomas, ed., *World Security: Trends and Challenges at Century's End* (New York: St. Martins, 1991), 401-424.

Kamieniecki, S. and **E. Sanasarian**. "Conducting Comparative Research on Environmental Policy." *Natural Resources Journal*, 30:2 (1990), 321-340.

Kay, D., and **E. Skolnikoff**, ed. *World Eco-Crisis: International Organizations in Response*. Wisconsin: University of Wisconsin Press, 1972.

Keeley, J. "The Latest Wave: A Critical Review of Regime Literature." In D. Haglund and M. Hawes, ed., *World Politics: Power, Interdependence and Dependence* (Toronto: Harcourt Brace Jovanovich, 1990), 553-569.

Keohane, R. *After Hegemony: Cooperation and Discord in the World Political Economy*. Princeton: Princeton University Press, 1984.

Kolodziej, E. "Renaissance in Security Studies? Caveat Lector!" *International Studies Quarterly*, 36 (1992), 421-438.

Kothari, R. "Peace in an Age of Transformation." In R.B.J. Walker, ed., *Culture, Ideology, and World Order* (Boulder: Westview, 1984), 323-361.

Krasner, S. "Structural Causes and Regime Consequences: Regimes as Intervening Variables." In same, ed., *International Regimes* (Ithica: Cornell University Press, 1983), 1-22.

Long, D. "From World Peace to International Organization: The Development of Functionalism As A Theory of International Relations." Paper presented to the Canadian Political Science Association Annual Meeting, Carleton University, Ottawa, 6-8 June 1993.

Lowe, P., and **W. Rudig.** "Review Article: Political Ecology and The Social Sciences -- The State of the Art." *British Journal of Political Science,* 16:4 (1986), 513-550.

Lunds, L. "North/South and Global Warming -- Conflict or Cooperation?" *Bulletin of Peace Proposals,* 22:2 (1991), 199-210.

Luttwak, E. "Interaction and Access to Natural Resources." In H. Bull, ed., *Intervention in World Politics* (Oxford: Clavendon, 1984), 79-94.

MacNeil, J. "The Greening of International Relations." *International Journal,* XLV (1989-90), 1-35.

——— et. al. *Beyond Interdependence: The Meshing of the World's Economy and the Earth's Ecology.* New York: Oxford University Press, 1991.

Mangold, P. "Security: New Ideas, Old Ambiguities." *The World Today* (Royal Institute of International Affairs), 47:2 (1991), 30-33.

Mann, D. "Environmental Learning in a Decentralized Political World." *Journal of International Affairs,* 44:2 (1991), 301-338.

Mansbach, R. "The World Turned Upside Down." *The Journal of East Asian Affairs,* 7:2 (1993), 451-497.

McDougal, M., and **J. Schneider.** "The Protection of the Environment and World Public Order: Some Recent Developments." *Mississippi Law Journal,* 45 (1974), 1085-1124.

Mische, P. "Ecological Security and the Need to Reconceptualize Sovereignty." *Alternatives,* XIV (1989), 389-429.

Mitrany, D. *A Working Peace System: An Argument for the Functional Development of International Organization.* London: RIIA, 1943.

——— *The Progress of International Government.* New Haven: Yale, 1933.

Morgenthau, H. *Politics Among Nations: The Struggle For Power and Peace.* New York: Knopf, 1948.

Myers, N. "Environment and Security." *Foreign Policy,* 74 (1989), 23-41.

Nadelmann, E. "Global Prohibition Regimes: The Evolution of Norms in International Society." *International Organization,* 44:4 (1990), 481-526.

Newman, D. "The New Diplomatic Agenda: Are Governments Ready?" *International Affairs,* 65:1 (1989), 29-42.

Ophuls, W. *Ecology and the Politics of Scarcity: Prologue to a Political Theory of the Steady State.* San Francisco: Freeman, 1977.

Orr, D. "Modernization and Conflict: The Second Image Implications of Scarcity." *International Studies Quarterly,* 21 (1977), 593-618.

Osterad, O. "The Uses and Abuses of Geopolitics." *Journal of Peace Research,* 25:2 (1988), 191-199.

Palme Commission on Disarmament and Security Issues. *A World at Peace: Common Security in the 21st Century.* Stockholm, 1989.

Pentland, C. "Integration, Interdependence, and Institutions." In D. Haglund and M. Hawes, ed., *World Politics: Power, Interdependence and Dependence* (Toronto: Harcourt Brace Jovanovich, 1990), 173-196.

Pirages, D. "The Ecological Perspective and the Social Sciences." *International Studies Quarterly,* 27 (1983), 243-255.

——— *Global Ecopolitics: The New Context for International Relations.* Massachusetts: Duxbury, 1978.

Prins, G. "Politics and the Environment." *International Affairs,* 66:4 (1990), 711-730.

Puchala, D. "The United Nations and Ecosystem Issues: Institutionalizing the Global Interest" in L. Finkelstein, ed., *Politics in the United Nations System* (Durham: Duke, 1988), 214-245.

Ramakrishna, K. "North-South Issues, the Common Heritage of Mankind and Global Environmental Change." In I. Rowlands and M. Greene, ed., *Global Environmental Change and International Relations* (London: Macmillan and Millennium, 1992), 145-168.

Renner, M. *National Security: The Economic and Environmental Dimensions.* WorldWatch Paper 89 (Washington, 1989).

Rosenau, J. *Turbulence in World Politics.* Princeton: Princeton University Press, 1990.

Sands, P. "EC Environmental Law: The Evolution of a Regional Regime of International Environmental Protection." *The Yale Law Journal,* 100:8 (1991), 2511-2523.

Schriver, N. "International Organization for Environmental Security." *Bulletin of Peace Proposals,* 20:2 (1989), 115-122.

Shaw, T. "Security Redefined: Unconventional Conflict in Africa." In S. Wright and J. Brownfoot, ed., *Africa in World Politics: Changing Perspectives* (London: Macmillan, 1987),17-34.

Shiva, V. *Staying Alive: Women, Ecology and Development.* London: Zed, 1988.

Siebert, H. "Environmental Policy and European Integration." In same, ed., *Environmental Scarcity: The International Dimension* (Schriftteitumlg: Hubertus Muller-Groeling, 1991), 57-70.

Simmons, I.G. *Changing the Face of the Earth: Culture, Environment, History.* Oxford: Basil Blackwell, 1989.

Sorensen, Theodore. "Rethinking National Security." *Foreign Affairs*, 69:3 (1990), 1-18.

Soroos, M. "The Commons and Lifeboat as Guides for International Ecological Policy." *International Studies Quarterly*, 21:4 (1977), 647-674.

Speth, J. "Environment, Economy, Security." In same, *Protecting Our Environment* (Washington: Center for National Policy, 1985).

Springer A. "Protecting the Environment: A New Focus for the Atlantic Alliance." *The New Europe: Revolution in East-West Relations*, Proceedings of the Academy of Political Science, 38:1 (NY, 1991), 129-139.

Sprout, H. and **M.** *Toward a Politics of the Planet Earth.* New York: van Nostrand Reinhold, 1971.

Stanley Foundation, "Report of the 24th United Nations of the Next Decade Conference: Environmental Problems -- A Global Security Threat." June 18-23, 1989, Hamilton Parish, Bermuda.

Stoett, P. "The Environmental Enlightenment: Security Analysis Meets Ecology." *Coexistence: A Review of East-West and Development Issues*, 31 (1994), 127-146.

——— "Competing Perspectives of World Politics: Environmental Crises and Global Security." Paper presented to the Fifth Conference on the United Nations and Peace, University of Victoria, British Columbia, Canada, Feb. 1992.

——— "Of Opportunity and Necessity: The UNEP and the Functionalist Approach." *International Insights*, 6:1 (1990), 47-54.

Strange, S. "*Cave! hic dragones:* A Critique of Regime Analysis." In S. Krasner, ed., *International Regimes* (Ithaca: Cornell University Press, 1983), 337-354.

Subrahmanyam, K. "Alternative Security Doctrines." *Bulletin of Peace Proposals*, 22:1 (1990), 77-85.

Suliman, M. "Civil War in the Sudan: From Ethnic to Ecological Conflict." *The Ecologist*, 23:3 (1993), 104-109.

Taylor, P. *International Organization in the Modern World: the Regional and the Global Processes.* London: Pinter, 1993.

Terriff, T. "The 'Earth Summit': Are There Any Security Implications?" Forthcoming in *Arms Control: Contemporary Security Policy*, 13:2 (1992).

Thomas, R. "Who is the Enemy? Redefining Security." *Ecodecision: Environmental Policy Magazine*, 9 (1993), 41-44.

Timberlake, L. and **J. Tinker.** "The Environmental Origins of Political Conflict." *Socialist Review*, 15:6 (1983).

Tuchman Mathews, J., ed. *Preserving the Global Environment: The Challenge of Shared Leadership.* New York: Norton, 1991.

——— "Redefining Security." *Foreign Affairs*, 68:2 (1989), 162-177.

Ullman, R. "Redefining Security." *International Security*, 8:1 (1983), 129-153.

Viotti, P., and **M. Kauppi.** *International Relations Theory: Realism, Pluralism, Globalism.* New York: Macmillan, 1992.

Walker, R.B.J. *Inside\Outside: International Relations as Political Theory.* Cambridge: C.U.P. 1992.

——— "Security, Sovereignty, and the Challenge of World Politics." *Alternatives*, XV (1990), 3-27.

———*One World, Many Worlds: Struggles for a Just World Peace.* London: Zed, 1988.

——— and **S. Mendlovitz**, ed. *Contending Sovereignties: Redefining Political Community.* Boulder: Lynn Reinner, 1990.

Wallerstein, I. *The Capitalist World Economy.* London: Cambridge University Press, 1979.

Ward, B. *Progress for a Small Planet.* Harmondsworth: Penguin, 1979.

Westing, A. "The Environmental Component of Comprehensive Security", *Bulletin of Peace Proposals*, 20 (1989), 129-134.

Welsh Brown, J., ed. *In the U.S. Interest: Resources, Growth, and Security in the Developing World.* Boulder: Westview, 1990.

Wilson, T. "World Security and the Global Agenda." In H. Cleveland, ed., *The Management of Sustainable Growth* (New York: Pergamon, 1981).

Wolfers, A. "National Security as an Ambiguous Symbol." *Political Science Quarterly*, LXVII:4 (1952), 481-502.

Yearly, S. *The Green Case: A Sociology of Environmental Issues, Arguments and Politics.* London: HarperCollinsAcademic, 1991.

Young, O. "The Politics of International Regime Formation: Managing Natural Resources and the Environment." *International Organization*, 43:3 (1989), 349-375.

——— *International Cooperation: Building Regimes for Natural Resources and the Environment.* Ithaca: Cornell University Press, 1989.

———*Resource Regimes: Natural Resources and Social Institutions.* Berkeley: University of California Press, 1982

Zacher, M. "Toward a Theory of International Regimes", *Journal of International Affairs*, 44:1 (1990), 139-158.

CHAPTER THREE: GES AND INTERNATIONAL ORGANIZATION

Ager, D. "Ore That Launched a Thousand Ships." *New Scientist*, 106:1461 (1985), 28-29.

Almond, H. "War, Weapons and the Environment." *The Georgetown International Environmental Law Review*, 1:2 (1988), 167-198.

Bush, K. "Climate Change, Global Security, and International Governance." *Canadian Institute for International Peace and Security Working Paper 23;* Proceedings of a conference held in Ottawa, 11-12 April 1990. Microlog 90-07058.

Canada, Government of. Tucker, C.M. *Environmentalism and Environmental Change as it Affects Canadian Defence Policy and Operations.* "Part Three: Arctic Issues"; Operational Research and Analysis Establishment Report No. R106 (Ottawa, 1990), Microlog 91-01753.9.

Cox, R. "Multilateralism and World Order." *Review of International Studies*, 18 (1992).

Biehl, J. "Ecofeminism and Deep Ecology: Unresolvable Conflict?" In D. Roussopoulas, ed., *The Anarchist Papers: 2* (Montreal: Black Rose, 1989), 19-31.

Boyden, S. *BioHistory: The Interplay Between Human Society and the Biosphere, Past and Present.* Paris: UNESCO, 1992.

Breuer, G. *Weather Modification: Prospects and Problems.* Trans. by Hans Morth. Cambridge University Press, 1979.

Brown, S. "Explaining the Transformation of World Politics." *International Journal*, XLVI:2 (1991), 207-219.

Brown Weiss, E. *In Fairness to Future Generations: International Law, Common Patrimony, and Intergenerational Equity.* New York: Transnational and UNU, 1989.

IBRD. **Dejene, A.** and **J. Olivares**. "Integrating Environmental Issues Into A Strategy For Sustainable Agricultural Development." *World Bank Technical Paper No. 146* (Washington, 1991).

Catton, W. *Overshoot: The Ecological Basis of Revolutionary Change.* Chicago: University of Illinois, 1982.

Choucri, Nazli. "Perspectives on Population and Conflict." In same, ed., *Multidisciplinary Perspectives on Population and Conflict* (Syracuse: Syracuse University Press, 1984); 1-25.

——— and **James Bennett.** "Population, Resources, and Technology: Political Implications of the Environmental Crisis." In D. Kay and E. Skolnikoff, eds., *World Eco-Crisis: International Organizations in Response* (University of Wisconsin, 1972); 9-46.

——— and **Robert North.** *Nations in Conflict.* San Francisco: W.H. Freeman, 1975.

Cline, William. *The Economics of Global Warming.* Washington: Institute for International Economics, 1992.

D'Amato, A. "Do We Owe a Duty to Future Generations to Preserve the Global Environment?" *American Journal of International Law,* 84 (1990).

Daudelin, J. "L'environnement comme Cheval de Troie? Le cas de l'Amazonie breesilienne." Paper presented to the Canadian Political Science Association, June 1993, Carleton University, Ottawa.

Devall, W., and **G. Sessions.** *Deep Ecology: Living As If Nature Mattered.* Salt Lake City: G.M. Smith, 1985.

Drengson, A. "Shifting Paradigms: From the Technocratic to the Person-Planetary." *Environmental Ethics,* 2 (1980), 221-240.

Durham, William. *Scarcity and Survival in Central America:Ecological Origins of the Soccer War.* Stanford University Press, 1979.

Fauteux, P. "The Use of the Environment As An Instrument of War in Occupied Kuwait." In B. Schiefer, ed., *Verifying Obligations Respecting Arms Control and The Environment: A Post Gulf War Assessment* (University of Saskatchewan, 1992), 35-79.

Feman, M.E. "Review and Impacts of Climate Change Uncertainties." *Futures: The Journal of Forecasting, Planning and Policy,* 25:8 (1993), 850-863.

Fornos, W. *Gaining People, Losing Ground.* Washington: The Population Institute, 1990.

Fre, Z. "The Legacy of War." In O. Bennett, ed., *Greenwar: Environment and Conflict* (London: Panos Institute, 1991), 131-142.

Freeman, Bill. *Environmental Ecology: The Impact of Pollution and Other Stresses on Ecosystem Structure and Function.* Academic Press, 1989.

Goldblat, Josef. "Legal Protection of the Environment Against the Effects of Military Activities." *Bulletin of Peace Proposals,* 22:4 (1991), 399-406.

Goldemberg, J., and **E. R. Durham.** "Amazonia and National Sovereignty." *International Environmental Affairs,* 2:1 (1990), 22-39.

Gribben, J. *The Hole in the Sky: Man's Threat to the Ozone Layer.* London: Corgi, 1988.

Gupta, R.S. "Resolutions of the UNGA as a Source of International Law." *International Studies,* 23:2 (1986), 143-154.

Gurr, T. "On the Political Consequences of Scarcity and Economic Decline." *International Studies Quarterly,* 29:1 (1985), 51-75.

Haglund, D. "The New Geopolitics of Minerals: An Inquiry Into the Changing International Significance of Strategic Minerals." In same, ed., *The New Geopolitics of Strategic Minerals: Canada and International Resource Trade* (Vancouver: University of B.C., 1989), 3-34.

Hammond, A., ed. *The 1993 Information Please Environmental Almanac, Compiled By the World Resources Institute.* Boston: Houghton Mifflin.

Hankins, F. "Pressure of Population as a Cause for War." *Annals of the American Academy of Political and Social Science,* 198:1 (1938), 101-108.

Herz, J. "Idealist Internationalism and the Security Dilemma." *World Politics,* 2 (1950), 157-180.

Jacobsen, H. *Networks of Interdependence:International Organisations and the Global Political System,* Second Ed. New York: Knopf, 1984.

Jorgensen, J. "Global Interdependence: Dimensions and Dilemmas." In H. Wiseman, et. al., ed., *Ethics and Technology: Ethical Choices in the Age of Pervasive Technology* (Toronto: Wall and Thompson, 1989), 172-179.

Juda, L. "Negotiating a Treaty on Environmental Warfare and Its Impact Upon Arms Control Negotiations." *International Organization,* 32:4 (1978).

Karpe, H.J., et. al., eds. *Climate and Development: Climatic Change and Variability and the Resulting Social, Economic, and Technological Implications.* Berlin: Springer-Verlag, 1990.

Krasner, S. *Structural Conflict: The Third World Against Global Liberalism.* Los Angeles: University of California, 1985.

Kratochwil, F. and **J. Ruggie,** "International Organisation: The State of the Art." In P. Diehl, ed., *The Politics of International Organisations: Patterns and Insights* (Chicago: Dorsey, 19), 17-27.

Laszlo, E. "Cooperative Governance", *World Futures,* 31 (1991), 215-221.

——— *The Systems View of the World.* New York: Braziller, 1972.

Leroy, M. *Population and World Politics: The Interrelationship Between Demographic Factors and International Relations.* Leiden: Netherlands Interuniversity Demographic Institute, 1978.

Lesser, I. *Resources and Strategy.* London: Macmillan, 1989.

Lipschutz, R. *When Nations Clash: Raw Materials, Ideology, and Foreign Policy*. New York: Ballinger, 1989.

Lovelock, J. *Gaia: A New Look at Life on Earth*. New York: Oxford University Press, 1979.

MacDonald, G. "How to Wreck the Environment." In N. Calder, ed., *Unless Peace Comes* (London: Penguin, 1968), 167-83.

Markham, S.F. *Climate and Energy of Nations*. London: Oxford, 1944.

McCormick, J. *Acid Earth: The Global Threat of Acid Pollution*. London: Earthscan, 1985.

Miller, M. *Debt and the Environment: Converging Crises*. New York: United Nations, 1991.

Mills, C.A. *World Power and Shifting Climate*. Boston: Christopher, 1963.

Minger, T., ed. *Greenhouse Glasnost: The Crisis of Global Warming*. New York: Ecco, 1990.

Munton, D. "Dependence and Interdependence in Transboundary Environmental Relations." *International Journal*, 36:1 (1981), 139-184.

Murphy, C. "The UN's Capacity to Promote Sustainable Development: The Lessons of a Year that 'Eludes All Facile Judgement.'" In *The State of The UN: 1992*, ACUNS Reports and Papers, No. 3 (1992), 49-75.

Myers, N. "Environment and Security." *Foreign Policy*, 74 (1989), 23-41.

Nelson, E. "Britain's Aboriginal Sin." *Bulletin of the Atomic Scientists*, 48:6 (1992), 8-9.

Oberg, J. "Ecological Ethics: Nature Has Rights, Humans Have Duties." *Bulletin of Peace Proposals* 21:4 (1990), 403-411.

Ophuls, W., and **A.S. Boyan.** *Ecology and the Politics of Scarcity Revisited: The Unraveling of the American Dream*. New York: Freeman, 1992.

O'Riordan, T. *Environmentalism*. London: Pion, 1976.

Oxman, B. "Environmental Warfare." *Ocean Development and International Law*, 22:4 (1992), 433-437.

Page, D. "Debt-for-Nature Swaps: Experience Gained, Lessons Learned." *International Environmental Affairs*, 1:4 (1989), 275-288.

Park, C. *Acid Rain: Rhetoric and Reality*. London: Methuen, 1987.

Pentland, C. "International Organisations and Their Roles." In P. Diehl, ed., *The Politics of International Organisations: Patterns and Insights* (Chicago: Dorsey, 1989), 5-14.

Renner, M. "Assessing the Military's War on the Environment." In L. Brown, ed., *State of the World 1991: A WorldWatch Institute Report on Progress Toward a Sustainable Society* (New York: Norton, 1992), 132-152.

Rice, R. "A Casualty of War: The Nicaraguan Environment." *Technology Review*, 92:4 (1989), 62-71.

Roots, E.F. "International Agreements to Prohibit or Control Modification of the Environment for Military Purposes: An Historical Overview and Comments on Current Issues." In B. Schiefer, ed., *Verifying Obligations Respecting Arms Control and The Environment: A Post Gulf War Assessment* (University of Saskatchewan, 1992), 13-34.

Rosenau, J. "Normative Challenges in a Turbulent World." *Ethics and International Affairs*, 6 (1992), 1-20.

Rusk, J. "Global Warming Could Melt Sovereignty." *The Globe and Mail*, July 21, 1992.

Russell, R. "The Nature of Military Impacts on the Environment." In Sierra Club, *Air, Water, Earth, Fire* (San Francisco: Sierra Club, 1974), 1-14.

Schmandt, J. et. al., ed. *Acid Rain and Friendly Neighbours: The Policy Dispute Between Canada and the United States*, Revised Edition. Duke University Press, 1988.

Schneider, S. *Global Warming: Are We Entering the Greenhouse Century?* San Francisco: Sierra Club, 1989.

Schreiber, H. "The Threat From Environmental Destruction in Eastern Europe." *Journal of International Affairs*, 44:2 (1991), 359-392.

Sebenius, J. "Designing Negotiations Toward a New Regime: The Case of Global Warming." *International Security*, 15:4 (1991), 110-148.

Shaw, R.P. "Rapid Population Growth and Environmental Degradation: Ultimate Versus Proximate Factors." *Environmental Conservation*, 16 (1989), 199-208.

Shiva, V. et. al. *Biodiversity: Social and Ecological Perspectives.* London: Zed, 1991.

Shoumatoff, A. *The World Is Burning.* Boston: Little, Brown and Company, 1990.

Simmons, I.G. *Changing the Face of the Earth: Culture, Environment, History.* Oxford: Basil Blackwell, 1989.

Simon, J. "Lebensraum: Paradoxically, Population Growth May Eventually End Wars." *Journal of Conflict Resolution*, 33:1 (1989), 164-180.

Soroos, M. "Ecology and the Time Dimension in Human Relationships." In same and D. Orr, ed., *The Global Predicament* (N.C.: North Carolina University Press, 1979), 337-343.

Springer, A. "International Environmental Law after Rio: The Continuing Search for Equity." *Ethics and International Relations*, 7 (1993), 115-130.

Stone, C. "The Law as a Force in Shaping Cultural Norms Relating to War and the Environment." In A. Westing, ed., *Cultural Norms, War and the Environment* (Oxford: Oxford University Press), 64-82.

Theorin, B. "Military Resources to the Environment." *Bulletin of Peace Proposals,* 23:2 (1992), 119-122.

UNEP. "The Effects of Weapons on Ecosystems." *United Nations Environment Programme Studies, Volume One,* by J.P. Robinson. Toronto: Permagon, 1979.

UNFPA. *Population, Resources and the Environment: The Critical Challenges.* Chief Writer: Norman Myers. London: Banson, 1991.

——— *Meeting the Population Challenge.* New York, 1989.

——— *Safeguarding the Future.* (Written by N. Sadik) New York: 1989.

UNIDR. "Disarmament, Environment, and Development and Their Relevance to the Least Developed Countries", Research Paper No. 10, A. Westing. (Geneva: UNIDR and UNEP, 1991).

VanDeVeer, D., and **C. Pierce,** ed. *People, Penguins, and Plastic Trees: Basic Issues in Environmental Ethics* (Belmont: Wadsworth, 1986).

Verwey, Wil D. *Riot Control Agents and Herbicides in War: Their Humanitarian, Toxicological, Ecological, Military, Polemological, and Legal Aspects.* Leydon: Sijthoff, 1977.

Weiner, M. "Security, Stability, and International Migration." *International Security,* 17:3 (1993), 91-126.

Weisberg, Barry, ed. *Ecocide in Indochina: The Ecology of War.* San Francisco: Canfield, 1970.

Westing, A. *Warfare in a Fragile World: Military Impact on the Human Environment.* London: Taylor and Francis, 1980.

———"The Environmental Impact of Conventional Warfare", in W. Barnaby, ed., *War and Environment* (Stockholm: Royal Ministry of Agriculture, 1981), 58-72.

——— *Environmental Warfare: A Technical, Legal and Policy Appraisal.* London: Taylor and Francis, 1984.

——— *Herbicides in War: The Long-term Ecological and Human Consequences.* London: Taylor and Francis, 1984.

———*Explosive Remnants of War: Mitigating the Environmental Effects.* Philadelphia: Taylor and Francis, 1985.

———*Global Resources and International Conflict: Environmental Factors in Strategic Policy and Action.* Oxford: Oxford, 1986.

——— *Comprehensive Security for the Baltic: An Environmental Approach.* London: Sage, 1989.

White, R. "Environmental Management and National Security: Some Issues From Senegal." *International Journal,* XLV (1990), 106-137.

Wigley, T., et al, eds. *Climate and History.* Cambridge: Cambridge University Press, 1981.

Wijkman, P. "Managing the Global Common." *International Organization,* 36:3 (1982), 511-536.

Williams, P. and **D. Wallace.** *Unit 731: Japan's Secret Biological Warfare in World War Two.* New York: The Free Press, 1989.

Wilson, E.O. *Biophilia: The Human Bond With Other Species.* London: Cambridge University Press, 1984.

Wirick, G. "Environment and Security: The Case of Central America." *Peace and Security* (Canadian Centre for International Peace and Security), Summer, 1989, 2-3.

Wirth, D. "Catastrophic Climate Change." In M. Klare and D. Thomas, ed., *World Security: Trends and Challenges at Century's End* (New York: St. Martin's, 1991), 381-400.

Woodwell, G. "The Effects of Global Warning." In J. Leggett, ed., *Global Warming: The Greenpeace Report* (Oxford: Oxford, 1990).

Wright, S., ed. *Preventing a Biological Arms Race.* Cambridge: MIT, 1990.

Yearley, S. *The Green Case: A Sociology of Environmental Issues, Arguments and Politics.* London: HarperCollinsAcademic, 1991.

Young, J. *Post Environmentalism.* London, 1990.

Young, J. "Mining the Earth", in L. Brown, ed., *State of the World: 1992* (New York: Norton, 1992), 100-118.

ZumBrunnen, C. "The Environmental Challenges in Eastern Europe." In I. Rowlands and M. Greene, ed., *Global Environmental Change and International Relations* (London: Macmillan, 1992), 88-121.

CHAPTER FOUR: ATOMS

Ahearne, J. "Fixing the Nation's Nuclear Weapons Plants." *Technology Review*, 92:5 (1989), 24-29.

Alvarez, R., and **A. Makhijani.** "Hidden Legacy of the Arms Race: Radioactive Waste." *Technology Review*, 91:6 (1988), 42-5.

Bernauer, T. "The End of Chemical Warfare." *Security Dialogue*, 24:1 (1993), 97-112.

Bertscht, G., and **R. Cupitt.** "Nonproliferation in the 1990s: Enhancing International Cooperation on Export Controls." *The Washington Quarterly*, 16:4 (1993), 53-70.

Blowers, A., **D. Lowry** and **B. Solomon.** *The International Politics of Nuclear Waste.* London: MacMillan, 1991.

Bujon de l'Estang, F. "The Delicate Balance: Governmental and Industrial Cooperation in Enforcing Nuclear Nonproliferation." In R. Jones, et. al., ed., *The Nuclear Suppliers and Nonproliferation: International Policy Choices* (Lexington: Lexington Books, 1985), 131-142.

Bukharin, O. "Soft Landing for Bomb Uranium." *The Bulletin of the Atomic Scientists*, 49:7 (1993), 44-49.

Butler, R. "Beyond the NPT." In J. Pilat and R. Pendley, eds., *Beyond 1995: The Future of the NPT Regime* (New York: Plenum, 1990), 65-71.

Camilleri, J. *The State and Nuclear Power: Conflict and Control in the Western World.* Seattle: University of Washington Press, 1984.

Caufield, C. *Multiple Exposures: Chronicles of the Radiation Age.* London: Secker and Warbug, 1989.

Cervenka, Z., and **B. Rogers.** *The Nuclear Axis: Secret Collaboration Between West Germany and South Africa* (London: J. Friedman, 1978).

Divine, R. *Blowing on the Wind: The Nuclear Test Ban Debate, 1934-1960.* New York: Oxford, 1978.

Dorn, W., and **A. Rolya.** "The Organization for the Prohibition of Chemical Weapons and the IAEA: A Comparative Overview." *IAEA Bulletin*, 35:3 (1993), 44-48.

Easlea, B. *Fathering the Unthinkable: Masculinity, Scientists and the Nuclear Arms Race.* London: Pluto, 1983.

Ehrlich, P., et. al. "Long-term Biological Consequences of Nuclear War." *Science*, 23 (1983), 1293-96.

Epstein, W. "The Linkage Between a Nuclear Test Ban and Nuclear Non-Proliferation." In M. Fry, et. al., eds., *Nuclear-Proliferation and the Non-Proliferation Treaty* (Berlin: Springer-Verlag, 1990), 132-146.

Falk, J. *Global Fission: The Battle Over Nuclear Power.* New York: Oxford University, 1982.

Finch, R. *Exporting Danger: A History of the Canadian Nuclear Energy Export Programme.* Montreal: Black Rose, 1986.

Frankel, B., ed. *Opaque Nuclear Proliferation: Methodological and Policy Implications.* London: Frank Cass, 1991.

Gallagher, C. *American Ground Zero: The Secret Nuclear War.* Cambridge, MA: MIT Press, 1993.

Gilinsky, V. "International Discipline Over the uses of Nuclear Energy." In A. Wohlstetter, et. al., eds., *Nuclear Policies: Fuel Without the Bomb* (Cambridge:Ballinger, 1978), 73-82.

Gizewski, P. "Military Activity and Environmental Security: The Case of the Arctic." Paper presented to the "Environmental Security After Communism" Conference, Carleton University, Ottawa, Feb. 26-27, 1993.

Goldblat, J. "The Nuclear Non-Proliferation Imperative." *Bulletin of Peace Proposals,* 17:3-4 (1986), 491-496.

Goldemberg, J. "Nonproliferation Nonsense." *New York Times,* Nov. 18, 1993, A19.

Goldschmidt, B., and **M. Kratzer.** "Peaceful Nuclear Relations: A Study of the Creation and the Erosion of Confidence." In I. Smart, ed., *World Nuclear Energy: Toward a Bargain of Confidence* (Baltimore: Johns Hopkins University Press, 1982), 19-48.

Gregory, S., and **A. Edwards.** "The Hidden Cost of Deterrence: Nuclear Weapons Accidents, 1950-1988." *Bulletin of Peace Proposals,* 20:1 (1989), 3-26.

Haig, Brian. "Why We Must Accept Nuclear Winter Theory." *Bulletin of Peace Proposals,* 20:1 (1989), 81-88.

Halverson, T. "Ticking Time Bombs: East Bloc Reactors." *The Bulletin of the Atomic Scientists,* July\August 1993.

Hansen, C. *U.S. Nuclear Weapons: The Secret History.* New York: Orion Books, 1988.

Hertsgaard, M. *Nuclear Inc.: The Men and Money Behind Nuclear Energy.* New York: Pantheon, 1983.

Hilgartner, S., et. al. *Nukespeak: The Selling of Nuclear Technology in America.* New York: Penguin, 1983.

Ivanova, T. and **N. Glovatskaia.** "Criteria and Principles of Developing an Ecological Policy", trans. by A. Schultz. *The Soviet Review,* 33:1 (1992), 32-44.

Jackson, J. "Nuclear Time Bombs." *Time,* December 7, 1992, 44- 49.

Janofsky, M. "Fallout Never Stops for Test Victims." *The Globe and Mail,* Jan. 12, 1994, A7.

Jonsson, C., and **S. Bolin,** "IAEA's Role in the International Politics of Atomic Energy." In L. Finkelstein, ed., *Politics in the United Nations System* (Durham: Duke University, 1988), 303-323.

Kapur, A. "Dump the Treaty." *Bulletin of Atomic Scientists,* July/August 1990, 21-22.

——— "World and Regional Power Relations Without the NPT." In J. Pilat and R. Pendley, eds., *Beyond 1995: The Future of the NPT Regime* (New York: Plenum, 1990), 117-131.

——— "Nuclear Energy, Nuclear Proliferation and National Security: Views From the South." In J. Keeley and R. Boardman, eds., *Nuclear Exports and World Politics: Policy and Regime* (New York: St. Martin's, 1983), 163-193.

Keeley, J. "Towards a Foucauldian Analysis of International Regimes." *International Organization,* 44:1 (1990).

Kennedy, R. "Nuclear Trade: Reliable Supply and Mutual Obligation." In R. Jones, et. al., ed.,*The Nuclear Suppliers and Nonproliferation: International Policy Choices* (Lexington: Lexington Books, 1985).

Khan, M.A. "Nuclear Energy and International Cooperation: A Third World Perspective of the Erosion of Confidence." In I. Smart, ed., *World Nuclear Energy: Toward a Bargain of Confidence* (Baltimore: Johns Hopkins University Press, 1982), 49-68.

Lane, E. "U.S., Russia Team Up for Nuclear Cleanup." *The Toronto Star,* May 22, D6, 1993.

Lenssen, N. "Facing Up to Nuclear Waste." *Worldwatch* March\April 1992 (Worldwatch Institute, Washington), 10-17.

List, M. "The West and Eastern Nuclear Power Plant Safety." *Aussen Politik,* 44:4 (1993), 344-353.

Lovins, A.B. and **L.H. Lovins.** *Energy\War: Breaking the Nuclear Link.* New York: Harper and Row, 1980.

MacKay, L. "Lines of Longitude: People's Detente, North and South." In M. Kalder and L. MacKay, eds., *Something in the Wind: Politics After Chernobyl* (London: Pluto, 1988), 131-160.

McMorris-Tate, T. "Regime Building in the Non-Proliferation System." *Journal of Peace Research,* 27:4 (1990), 399-414.

Mikesell, R. *Economic Development and the Environment: A Comparison of Sustainable Development with Conventional Development Economics.* London: Mansell, 1992.

Myers, S., et. al. "Prospects for the Power Sector in Nine Developing Countries." *Energy Policy*, 21:11 (1993), 1123-1132.

Novick, S. "The Menace of the Peaceful Atom." In F. Knelman, ed., *1984 and All That: Modern Science, Social Change, and Human Values* (Belmont: Wadsworth, 1972), 217-227.

Patterson, W. *Plutonium Business and the Spread of the Bomb*. London: 1984.

Peterson, J., ed. *The Aftermath: The Human and Ecological Consequences of Nuclear War*. New York: Pantheon, 1983.

Pilat, J. and **R. Pendley**, eds. *Beyond 1995: The Future of the NPT Regime*. New York: Plenum, 1990.

Power, P. "The Mixed State of Non-Proliferation: The NPT Review Conference and Beyond." *International Affairs*, 62:3 (1986), 477-491.

Preston, R., and **S. Wise.** *Men In Arms: A History of Warfare and Its Interrelationships with Western Society*, Fourth Edition. New York: Holt, Rinehart and Winston, 1979.

Price, T. "Uranium Supply and Demand." In E. Meller, ed., *Internationalization: An Alternative to Nuclear Proliferation?* (Cambridge, Mass.: Oelgeschlager, 1980), 51-70.

Prins, G., et. al. "The Case Against a Nuclear Defence Policy Part II: Nuclear Projects and the Environment." In same, *Defended to Death: A Study of the Nuclear Arms Race* (Harmondsworth: Penguin, 1983), 233-261.

Ramberg, B. "Targeting Nuclear Energy." In D. Ball and J. Richelson, ed., *Strategic Nuclear Targeting* (Ithaca: Cornell, 1986), 250-66.

Ramesh, T. "The UN in a Changing World." *Security Dialogue*, 24:1 (1993), 7-20.

Ray, A. "Third World Perspectives on Security and Nuclear Non-Proliferation: An Indian View." In J. Simpson, ed., *Nuclear Non-Proliferation: An Agenda for the 1990s* (Cambridge, New York: Cambridge University Press, 1987), 52-63.

Roberts, B. "From Nonproliferation to Antiproliferation." *International Security*, 18:1 (1993), 139-173.

Rueter, T. and **T. Kalil.** "Nuclear Strategy and Nuclear Winter." *World Politics*, 43 (1991), 587-607.

Sagan, C. "Nuclear War and Climatic Catastrophe: Some Policy Implications." *Foreign Affairs*, 62 (1984), 257-292.

Sagan, S. *The Limits of Safety: Organizations, Accidents, and Nuclear Weapons*. Princeton: Princeton University Press, 1993.

Sanders, B. "Non-Proliferation Diplomacy: Preparing For 1995." *Security Dialogue*, 24:2 (1993), 221-231.

Scheinman, L. *The International Atomic Energy Agency and World Nuclear Order*. Washington: Resources for the Future, 1987.

Schiff, B. *International Nuclear Technology Transfer: Dilemmas of Dissemination and Control.* London: 1983.

Seaborg, G., and **B. Loeb.** *Stemming the Tide: Arms Control in the Johnson Years*. Toronto: Lexington Books, 1987.

Semenov, B., and **N. Oi.** "Nuclear Fuel Cycles: Adjusting to New Realities." *IAEA Bulletin,* 35:3 (1993), 4.

Shulman, S. *The Threat at Home: Confronting the Toxic Legacy of the U.S. Military*. Boston: Beacon, 1992.

Simpson, J., and **D. Howlett.** "Nuclear Nonproliferation: The Way Forward." *Survival,* 33:6 (1991), 483-499.

Solingen, E. "Brazil: Technology, Countertrade, and Nuclear Exports." In W. Potter, eds., *International Nuclear Trade and Nonproliferation: the Challenge of the Emerging Suppliers*. (Lexington, Mass.: Lexington Books, 1990), 111-152.

Solomon, B. "The Politics of Nuclear Power and Radioactive Waste Disposal: From State Coercion to Procedural Justice?" *Political Geography Quarterly*, 7:3 (1988), 291-8.

Spector, L. and **J. Smith.** "Deadlock Damages Nonproliferation." *Bulletin of the Atomic Scientists,* Dec. 1990, 39-44.

Spence, M. "Exploring the 'Peaceful Atom': The Pattern of Nuclear Proliferation." *Radical Science,* 14 (1984), 80-93.

Steele, Karen. "Cleaning up After the Cold War." In World Resources Institute, *The 1993 Information Please Environmental Almanac* (Washington: WRI, 1993), 129-146.

Stoler, P. *Decline and Fall: The Ailing Nuclear Power Industry.* New York: Dodd, 1985.

Subrahmanyan, K. "Export Controls and the North-South Controversy." *The Washington Quarterly,* 16:2 (1993), 135-144.

Turco, R.P., et. al. "Nuclear Winter: Global Consequences of Multiple Nuclear Explosions." *Science,* 23, Dec. 1983, 1283-1292.

United Nations. *Study on the Climatic and Other Effects of Nuclear War. UN Report A/43/351.*

Vandercook, W. F. "Making the Very Best of the Very Worst: The 'Human Effects of Nuclear Weapons' Report of 1956." *International Security,* 11:1 (1986), 184-195.

Walker, J. "A Green Peace: The Closure of Overseas Military Installations and the Environmental Liability of the U.S. Government." *Georgetown International Environmental Law Review,* 3:2 (1990), 437-456.

Walker, W. "Nuclear Non-Proliferation and the Present Status of the International Nuclear Industry." In J. Simpson and A. McGraw, eds., *The International Nuclear Non-Proliferation System: Challenges and Choices* (New York: St. Martin's, 1984), 91-107.

——— and **D. Albright**, and **F. Berkhout.** *World Inventory of Plutonium of Highly Enriched Uranium.* (SIPRI: Oxford, 1993).

——— and **M. Lonnroth.** *Nuclear Power Struggles: Industrial Competition and Proliferation Control.* London, 1983.

Waltz, K. "The Spread of Nuclear Weapons: More May Be Better." *Adelphi Paper,* No. 171 (London: IISS, 1981).

Westing, A. *Weapons of Mass Destruction and the Environment.* London: Taylor and Francis, 1977.

Wilmhurst, M. "The Adequacy of IAEA Safeguards For the 1990s." In M. Fry, et. al., eds., *Nuclear-Proliferation and the Non-Proliferation Treaty* (Berlin: Springer-Verlag, 1990), 13-18.

Wonder, E. "The Nuclear Marketplace and the Nonproliferation Regime." In J. Pilat, ed., *The Nonproliferation Predicament* (New Brunswick, New Jersey: Transaction, 1985), 71-82.

Worsely, P., and **K. Buenor Hadjor,** eds. *On the Brink: Nuclear Proliferation and the Third World.* London: Third World Review, 1987.

CHAPTER FIVE: WHALES

Adams, J.S., and **T. McShane.** *The Myth of Wild Africa: Conservation Without Illusion.* New York: Norton, 1992.

Attfield, R. "Attitudes to Wildlife in the History of Ideas." *Environmental History Review,* 15: 2 (1991), 71-78.

Birnie, P. "The Role of Developing Countries in Nudging the International Whaling Commission From Regulating Whaling to Encouraging Non-consumptive Uses of Whales." *Ecology Law Quarterly,* 12 (1985), 937-975.

——— *Legal Measures for the Prevention of "Pirate" Whaling.* Switzerland: IUCN, 1981.

Boardman, R. *International Organisation and the Conservation of Nature.* Bloomington: Indiana University, 1981.

Bockstoce, J.R. *Whales, Ice, and Men: The History of Whaling in the Western Arctic.* Seattle: University of Washington Press, 1986.

Bureau of International Whaling Statistics. *International Whaling Statistics.* Oslo: Various years.

Burke, T. "Friends of the Earth and the Conservation of Resources." In P. Willetts, ed., *Pressure Groups in the Global System: The Transnational Relations of Issue-Oriented Non-Governmental Organisations* (London: Francis Pinter, 1982), 105-124.

Caldwell, L. "International Conservation Efforts: Cooperation to Protect Endangered Species and Environments." In same, *In Defense of the Earth: International Protection of the Biosphere* (Bloomington: Indiana University, 1972), 53-9O.

Caron, D. "International Sanctions, Ocean Management, and the Law of the Sea: A Study of Denial of Access to Fisheries." *Ecological Law Quarterly*, 16:1 (1989), 311-354.

Charney, J., ed. *The New Nationalism and the Use of Common Spaces: Issues in Marine Pollution and the Exploitation of Antarctica.* New Jersey: Allanheld, 1982.

Cherfas, J. *The Hunting of the Whale: A Tragedy That Must End.* London: Bodley Head, 1988.

Clark, C. "The Economics of Overexploitation." *Science*, 181:4100 (1973), 630-634.

Cook, J. *Pursuing the Whale: A Quarter-Century of Whaling in the Arctic.* London: John Murray, 1926.

Cousteau, J. and **P. Diole.** *The Whale: Mighty Monarch of the Sea.* New York: Doubleday, 1972.

D'Amato, A., and **S. Chopra.** "Whales: Their Emerging Right to Life." *American Journal of International Law* 85:1 (1991), 21-62.

Day, D. *The Whale War.* Vancouver: Douglas and McIntire, 1987.

Dorsey, K. "Putting a Ceiling on Sealing: Conservation and Cooperation in the International Arena, 1909-1911." *Environmental History Review*, 15:3 (1991), 27-46.

Doubleday, N. "Aboriginal Subsistence Whaling: The Right of Inuit to Hunt Whales and Implications for International Environmental Law." *Denver Journal of International law and Policy*, 17:2 (1989), 373-393.

Ehrenfeld, D. *Conserving Life on Earth.* New York: Oxford University Press, 1972.

Ehrlich, P., and **A. Ehrlich.** *Extinction: The Causes and Consequences of the Disappearance of Species.* New York: Random House, 1981.

Fichtelius, K., and **S. Sjolander.** *Man's Place: Intelligence in Whales, Dolphins, and Humans.* London: Gollancz, 1973.

Forster, M. "IWC: Trying Hard?" *Environmental Policy and Law,* 15 (1985), 8-9.

Francis, D. *Arctic Chase: A History of Whaling in Canada's North.* Toronto: Breakwater, 1984.

Glassner, M. *Neptune's Domain: A Political Geography of the Sea.* Boston: Unwin Hyman, 1990.

Haas, E. *When Knowledge Is Power: Three Models of Change in International Organisation.* Berkeley: University of California, 1990.

Harrison, R. "History of Whaling." In same and M. Bryden, ed., *Whales, Dolphins and Porpoises* (London: Merehurst, 1988), 182-195.

Haskell, R. "Abandoning Recent Whale Conservation Initiatives in Japan Whaling Association v. American Catacean Society." *Harvard Environmental Law Review,* 11:2 (1987), 551-591.

Hayden, S. *The International Protection of Wildlife: An Examination of Treaties and Other Agreements for the Preservation of Birds and Mammals.* New York: 1942.

Heazle, M. "A Community on Edge: Ban on Whaling Threatens Future of Ayukawa." *Far Eastern Economic Review,* 10 June 1993, 34-35.

Hunter, R., and **R. Weyler.** *To Save a Whale: The Voyages of Greenpeace.* Vancouver: Douglas and McIntyre, 1978.

IWC. *Chairman's Report of IWC Annual Meeting.* Cambridge: Various years.

Jackson, G. *The British Whaling Trade.* London: Adam and Charles Nlack, 1978.

Johnston, D.M. "The Driftnetting Problem in the Pacific Ocean: Legal Considerations and Diplomatic Options." *Ocean Development and International Law,* 21:1 (1990), 5-40.

Kaufman, L. "Why the Ark is Sinking." In same and K. Mallory, ed., *The Last Extinction* (Cambridge: MIT, 1986).

Kent, G. "Ocean Fisheries Management." In D. Orr and M. Soroos, ed., *Ecological Perspectives on World Order* (University of North Carolina, 1979), 232-248.

Knox, G. "The Key Role of Krill in the Ecosystem of the Southern Ocean With Special Reference to the Convention on the Conservation of Antarctic Marine Living Resources." *Ocean Management,* 9 (1984), 113-156.

Leornard, L. "The International Protection of Whales." *American Journal of International Law,* 35 (1941), 90.

Lyster, S. *International Wildlife Law: An Analysis of International Treaties Concerned With the Conservation of Wildlife.* Cambridge: Grotius, 1985.

Macgregor, J. "The Paradoxes of Wildlife Conservation in Africa." *Africa Insight,* 19:4 (1989), 201-212.

Mandel, Robert. "Transnational Resource Conflict: The Politics of Whaling." *International Studies Quarterly* 24:1 (1980); 99-127.

Manes, C. *Green Rage: Radical Environmentalism and the Making of Civilization.* Boston: Little and Brown, 1990.

McHugh, H. "The Role and History of the International Whaling Commission." In W. Schevill, ed.,*The Whale Problem: A Status Report* (Cambridge: Harvard, 1974), 305-335.

McNeely, J., et. al. *Conserving the World's Biological Diversity.* Washington: IUCN, 1990.

——— *Economics and Biological Diversity: Developing and Using Economic Incentives to Conserve Biological Resources.* Switzerland: IUCN, 1988.

M'Gonigle, R. "Economics, Ecology, and the Endangered Whale: Competing Foundations For Environmental Law and Politics." *Ecology Law Quarterly,* 9:1 (1980), 120-237.

Mowat, F. *A Whale For the Killing.* Toronto: McClelland and Stewart, 1972.

Myers, N. *The Sinking Ark: A New Look at the Problem of Disappearing Species.* Oxford: Pergamon, 1979.

Norton, B. *Why Preserve Natural Variety?* Princeton: Princeton University Press, 1987.

Paterson, D. *The Deluge and the Ark: A Journey Into Primate Worlds.* Boston: Houghton Mifflin, 1989.

Peterson, M. "Whalers, Cetologists, Environmentalists, and the International Management of Whaling." *International Organization,* 46:1 (1992), 147-186.

Pollack, A. "Commission to Save Whales Endangered, Too." *New York Times,* May 18, 1993, C4.

Ray, J. "The Abolition of Slavery and the End of International War", *International Organization,* 43:4 (1989), 406-439.

Sagoff, M. "On Preserving the Natural Environment." *Yale Law Journal,* 81 (1974), 205-267.

Schevill, W., ed. *The Whale Problem: A Status Report.* Cambridge: Harvard University Press, 1974.

Small, G. *The Blue Whale.* New York: Columbia, 1971.

Stackpole, E. *Whales and Destiny: The Rivalry Between America, France, And Britain for Control of the Southern Whale Fishery, 1785-1825.* Amherst: University of Massachusetts, 1972.

Stoett, P. "International Politics and the Protection of Great Whales." *Environmental Politics,* 2:2 (1993), 277-302.

-- "The IWC: Holding Firm on Whaling." *Sea Wind: Bulletin of Ocean Voice international,* 7:1 (1993), 22-23.

-- "International Organization and Normative Transition: The Death of Whaling and the Survival of the Whale." Paper presented to the Canadian Political Science Association, Prince Edward Island University, June 1992.

Swaney, J., and **P. Olson.** "The Economics of Biodiversity: Lives and Lifestyles." *Journal of Economic Issues,* XXVI:1 (1992), 1-26.

Teclaff, L. and **A. Utton,** ed. *International Environmental Law.* New York: Praeger, 1974.

Tonnessen, J., and **A. Johnsen.** *The History of Modern Whaling.* Trans. by R. Christophersen. Los Angeles: University of California Press, 1982.

van der Woude, R., and **A. Van Wijngaarden.** *Animals in Danger: A Study of Certain Mammal Species Threatened with Extinction in Europe.* Council of Europe, 1969.

von Moltke, K. "International Commissions and Implementation of International Law" in J. Carroll, ed., *International Environmental Diplomacy: The Management and Resolution of Transfrontier Environmental Problems* (New York: Cambridge, 1988), 87-93.

Watt, K. *Ecology and Resource Management.* New York: McGraw-Hill, 1968.

Webb, R.L. *On the Northwest: Commercial Whaling in the Pacific Northwest, 1790-1967.* Vancouver: UBC, 1988.

Weisman, S. "Whale: Food For Deep Thought, or Just Food?" *New York Times,* April 4, 1992, A6.

Westing, A. "Biodiversity and the Challenge of National Borders", *Environmental Conservation,* 20:1 (1993), 5-6.

Wilkinson, D. "The Use of Domestic Measures to Enforce International Whaling Agreements: A Critical Perspective." *Denver Journal of International Law and Policy,* 77:2 (1989), 271-292.

Williams, J., and **R. Nowak.** "Vanishing Species in Our Own Backyard: Extinct Fish and Wildlife of the United States and Canada." In J. Kaufman and K. Mallory, ed., *The Last Extinction* (Cambridge: MIT, 1986), 107-139.

Yutaka, H. "The Whaling Industry in Japan's Economy." In J. Schmidhauser and G. Totten, ed., *The Whaling Issue in U.S.-Japan Relations* (Boulder: Westview, 1978), 82-114.

CHAPTER SIX: RIVERS

Abuza, Z. "The Khmer Rouge Quest For Economic Independence." *Asian Survey: A Monthly Review of Contemporary Asian Affairs,* 33:10 (1993), 1010-1021.

Adams, P., and **L. Solomon.** *In the Name of Progress: The Underside of Foreign Aid,* Second Edition. Toronto: Energy Probe Research Foundation, 1992.

Arad, R. W., et. al. *Sharing Global Resources.* New York: McGraw-Hill, 1979.

Baltscheffsky, S. "Water Conflicts Set Nation Against Nation." *ENVIRO,* 15 (1993), 31-32.

Barber, M., and **G. Ryder,** ed. *Damming the Three Gorges: What Dam Builder's Don't Want You to Know (A Critique of the Three Gorges Water Control Feasibility Study),* Second Edition. Toronto: Probe International and Earthscan, 1993.

Braatz, S. *Conserving Biological Diversity: A Strategy for Protected Areas in the Asia-Pacific Region.* World Bank Technical Paper No. 193, Asia Technical Development Series. Washington: IBRD, 1992.

Broad, R., et. al. "Sustainable Development in the 1990s." In C. Hartman and P. Vilanova, ed., *Paradigms Lost: The Post Cold War Era* (London: Pluto, 1992), 89-110.

Burger, J. *Report From the Frontier: The State of the World's Indigenous Peoples.* London: Zed, 1987.

Cowell, A. "Hurdle to Peace: Parting the Mideast's Waters." *New York Times,* Oct. 10, 1993, 1.

Deudney, D. "The Mirage of Eco-War: The Weak Relationship Among Global Environmental Change, National Security and Interstate Violence." In I. Rowlands and M. Greene, *Global Environmental Change and International Relations* (London: Macmillan, 1992), 169-191.

Falkenmark, M. "Fresh Waters as a Factor in Strategic Policy and Action." In A. Westing, ed., *Global Resources and International Conflict* (Oxford: Oxford University Press, 1986), 85-113.

Frederiksen, H. *Water Resources Institutions: Some Principles and Practices.* World Bank Technical Paper No. 191. Washington: IBRD, 1992.

Galambos, J. "Political Aspects of an Environmental Conflict: The Case of the Gabcikovo-Nagymaros Dam System." In J. Kakoneen, ed., *Perspectives on Environmental Conflict and International Politics* (London: Pinter, 1992), 72-95.

Ghee, L. T., and **M. Valencia,** ed. *Conflict Over Natural Resources in South-East Asia and the Pacific.* Singapore: UNU and Oxford, 1990.

Gleick, P. "Water and Conflict: Fresh Water Resources and International Security." *International Security,* 18:1 (1993), 79-112.

——— *Water in Crisis.* New York: Oxford University Press, 1993.

——— "Climate Changes, International Rivers, and International Security: The Nile and the Colorado." In T. Minger, ed., *Greenhouse Glasnost: The Crisis of Global Warming* (New York: Ecco, 1990), 147-165.

Guimaraes, R. *The Ecopolitics of Development in the Third World: Politics and Environment in Brazil.* Boulder: Rienner, 1991.

Handley, P. "River of Promise." *Far Eastern Economic Review,* 16 Sept. 1993, 68-72.

Hiebert, M. "The Common Stream." *Far Eastern Economic Review,* Feb 21, 1991, 24-28.

——— and **P. Handley.** "Hostile Undercurrents." *Far Eastern Economic Review,* April 2, 1992, 16.

Hildyard, N., and **E. Goldsmith.** *The Social and Environmental Effects of large Dams,* Volume One and Two. Cornwall: Wadebridge Ecological Centre, 1984 and 1986.

Hubbel, D. "Antithetical Perceptions of Environment and Development: Village People and the State in Rural Thailand." Master in Environmental Studies Thesis, Faculty of Environmental Studies, York University, August 1992.

Hutton, W. "Chinese Giant Gets Ready to Shake the World." *The Globe and Mail,* Dec. 30, 1993, A21.

ICCILMB. *Mekong News: Newsletter of the Interim Mekong Committee.* Bangkok: Public Information Unit of the Mekong Secretariat, Various Editions.

IUCN. *Protected Areas and Demographic Change: Planning for the Future.* IVth World Congress on National Parks and Protected Areas, Caracas, February 1992. Gland, Switzerland, 1992.

Kakonen, J., ed. *Perspectives on Environmental Conflict and International Politics.* London: Pinter, 1992.

Kitazawa, Y. "The Japanese Economy and South-East Asia: The Examples of the Asahan Aluminium and Kawasaki Steel Projects." In L. Ghee and M. Valencia, eds., *Conflict Over Natural Resources in South-East Asia and the Pacific* (Singapore: UNU and Oxford, 1990), 51-93.

Kristof, N. "The Rise of China." *Foreign Affairs,* 72:5 (1993), 59-74.

Lele, S.M. "Sustainable Development: A Critical Review." *World Development,* 19:6 (1991), 607-622.

Lester, A.P. "River Pollution in International Law." *American Journal of International Law,* 57 (1963), 828-853.

Lightfoot, L. "Alternative Resettlement Strategies in Thailand: Lessons From Experience." In P. Gosling, ed., *Population Resettlement in the Mekong River Basin: Papers of the Third Carolina Geographical Symposium* (Chapel Hill: University of North Carolina, 1979), 28-38.

Lipschutz, R., and **J. Holdren.** "Crossing Borders: Resource Flows, the Global Environment, and International Security." *Bulletin of Peace Proposals,* 20:2 (1989); 123-128.

Lowi, M. "Bridging the Divide: Transboundary Resource Disputes and the Case of West Bank Water." *International Security,* 18:1 (1993), 113-138.

McBeth, J. "Mekong, No River of Peace." *Far Eastern Economic Review,* January 16, 1986, 28.

Mckenna, B. "Utilities Join Power to Tap Asian Demand", *The Globe and Mail,* October 8, 1993, B1.

MDRN, Coordination Office. "A Project Description of the Mekong Development Research Network." Institute of Asian Studies, Chulalongkorn, Bangkok: Phase I Application, Submitted to IDRC, July, 1992.

Mekong Secretariat and the Interim Mekong Committee. *Annual Report.* Bangkok: various years.

——— *Mekong News.* Bangkok: various editions.

——— "Cooperation in the Lower Mekong River Basin." In UN Department of Technical Cooperation for Development, *Experiences in the Development and management of International River and lake Basins,* Natural Resources / Water Series No. 10 (New York: UN, 1983), 245-251.

Mikesell, R. *Economic Development and the Environment: A Comparison of Sustainable Development with Conventional Development Economics.* London: Mansell, 1992.

Myers, S., et. al., "Prospects for the Power Sector in Nine Developing Countries", *Energy Policy,* 21:11 (1993), 1123-1132.

Nguyen Huu Ninh, et. al. "Energy in Vietnam." UN Centre for Science and Technology for Development, Advanced Technology Assessment System, No. 6, *Energy Systems, Environment and Development* (New York: 1991), 37-38.

Pantulu, V.B. "A Case-Study of the Pa Mong Project: Environmental Aspects." In UNEP, *Environment and Development in Asia and the Pacific: Experience and Prospects,* UNEP Reports and Proceedings Series No. 6, 1982, 145-159.

Rogers, P. "Economic and Institutional Issues: International River basins." In G. Le Moigre, et. al., eds., *Country Experiences with Water Resources Management: Economic, Institutional, Technological and Environmental Issues,* World Bank Technical paper No. 175 (Washington: IBRD, 1992), 63-70.

Said, A. A. *Human Rights and World Order.* New York: Praeger, 1978.

Savasdibutr, P. "The Development of the lower Mekong River Basin." In UN Department of Technical Co-operation for Development, *River and Lake Basin Development,* Natural Resources / Water Series, 20 (1990), 170-179.

Schaaf, C.H., and **R. Fifield.** *The Lower Mekong: Challenge to Cooperation in Southeast Asia.* Toronto: Van Nostrand, 1963.

Schramm, G. "Issues and Problems in the Power Sectors of Developing Countries." *Energy Policy,* 21:7 (1993), 735-747.

Shenon, P. "Thailand Stirs Criticism Over Cambodian Timber." *New York Times,* May 2, 1993, 21.

Sluiter, L. *The Mekong Currency.* Bangkok: TERRA, 1992.

Soroos, M. "Conflict in the Use and Management of International Commons." In J. Kakonen, ed., *Perspectives on Environmental Conflict and International Relations* (London: Pinter, 1992), 31-43.

Springer, A. "International Environmental Law After Rio: The Continuing Search for Equity." *Ethics and International Affairs,* 7 (1993), 115-130.

Starr, J. *U.S. Foreign Policy on Water Resources in the Middle East.* Washington: Center For Strategic and International Studies, 1987.

——— "Water Wars." *Foreign Policy,* 82 (Spring 1991), 17-36.

Stevens, R. *The Trail: A History of the Ho Chi Minh Trail and the Role of Nature in the War in Vietnam* (New York: Garland, 1993).

Stevens, W. "Why the Waters Must Flow Free." *The Globe and Mail,* July 21, 1993, A9.

Stoett, P. "Saving a Sick Paradigm or Articulating a New One?" *Futures: The Journal of Forecasting, Planning and Policy,* December 1993, 1,111-1,112.

——— "Regional International Organization: Southeast Asia's Mekong River Politics." Paper presented to the "Asia in the 1990s: Meeting and Making the New World" Conference, Queen's University, Kingston, October 1993.

Tasker, R. "Last Refuge." *Far Eastern Economic Review,* 16 Sept. 1993, 27.

Tefft, S. "Southeast Asians Face Off Over Mekong Dam Plan." *Christian Science Monitor,* May 21, 1993, 4.

TERRA. *Development in the Lower Mekong River Basin: A Briefing.* September 6, 1993.

——— Letter to C. Lankester, Executive Agent, Mekong Secretariat, Feb. 17, 1992.

Trolldalen, J. M. *International Environmental Conflict Resolution: The Role of the United Nations.* Washington: World Foundation for Environment and Development, 1992.

UNEP. *Environment and Development in Asia and the Pacific: Experience and Prospects.* UNEP Reports and Proceedings Sees, No. 6, Nairobi, 1992.

——— *Sustainable Development: Constraints and Opportunities.* New York: 1982.

——— *The Environmental Impacts of Production and Use of Energy: An Assessment Prepared By the UNEP.* Nairobi: Tycooly, 1981.

UNESCAP. "Sustainable and Environmentally Sound Development of the Water Resources in Thailand." In same, *Towards an Environmentally Sound and Sustainable Development of Water Resources in Asia and the Pacific,* Water Resources Series No. 71 (New York: UN, 1992), 201-211.

UNILC. *Law of the Non-Navigational Uses of International Watercourses.* Report of the ILC on the Work of its 43rd Session. New York: UN, 1991.

Valencia, J. "International Conflict Over Marine Resources in South-East Asia: Trends in Politicization and Militarization." In same and L.T. Ghee, eds., *Conflict Over Natural Resources in South-East Asia and the Pacific* (Singapore: Oxford/UN University, 1990), 94-144.

Vincent, R.J. *Foreign Policy and Human Rights: Issues and Responses.* Cambridge: Cambridge University, 1986.

Visser, F. "Recent Developments in the Joint Management of International Non-Maritime Water Resources in Africa." *The Comparative and International Law Journal of Southern Africa,* XXII:1 (1989); 59-92.

Wheeler, C. "The Mekong: From Tibet to the South China Sea." *IDRC Reports,* 21: 1 (April 1993), 26-27.

CHAPTER SEVEN: THE PARENTS OF INVENTION

Aikin, A.M. "Disposal or Uses of Large Quantities of Fission Products." *AECL Report on Atomic Power Symposium,* held at Chalk River, Ontario, 9-11 Sept. 1953 (CRR-548-A).

Boyden, S. *BioHistory: The Interplay Between Human Society and the Biosphere, Past and Present.* Paris: UNESCO, 1992.

Haas, E. *When Knowledge is Power: Three Models of Change in International Organization.* Berkeley: University of California Press, 1990.

Hoffmann, S. "An American Social Science: International Relations." *Daedalas,* 106 (1977).

Kennan, G. "To Prevent a World Wasteland: A Proposal." *Foreign Affairs,* 48 (1970), 401-413.

Marshal-Cornwall, J.H. *Geographic Disarmament: A Study of Regional Demilitarization.* London: Oxford, 1935.

Norse, E., ed. *Global Marine Biological Diversity: A Strategy for Building Conservation Into Decision Making.*Washington: Center for Marine Conservation, 1993.

PERSONAL CORRESPONDENCE

This book has benefited greatly from interviews, telephone conversations, or correspondence by mail or fax, from a wide variety of people outside those from the Queen's University community, including:

Robert Applebaum, Director-General of the International Directorate of Fisheries and Oceans, Government of Canada, Ottawa, January 1994.

Margaret Barber, Probe International, Toronto, August 1993.

Jim Beckett, Director, Fisheries Research, Government of Canada, Ottawa, January 1994.

Nathalie Billens, Secretary, Social Policy Service, IUCN, Switzerland, August 1993.

Robert Boardman, Professor of Political Science, Dalhousie University, Halifax, April 1994

Mary Caron, Education Coordinator, Population Institute, Washington, July 1993.

Simon Dalby, Assistant Professor, Dept. of Geography, Carleton University, August 1993.

Barbara Davidson, Information Officer, IDRC, Ottawa, July 1993.

Gloria Davis, Chief, Social Policy and Resettlement, Environment Department, IBRD, Washington, July 1993.

Ariel Delouya, Foreign Service Officer, Environment Division, Department of External Affairs, Government of Canada, Ottawa, July 1993.

Gerald Dirks, Associate Professor, Department of Political Studies, Brock University, St. Catherines, January, 1994

Simon Fairlie, Co-editor, *The Ecologist*, London, June 1993.

Qazi Shaukat Fareed, Director, Department of Humanitarian Affairs, United Nations, New York, July 1993.

Rachel Fleishman, International Affairs Officer, Office of the Deputy Under Secretary of Defense for Environmental Security, United States Government, Washington, July 1993.

Ray Gambell, Secretary, IWC, Cambridge, England, numerous occasions.

Peter Gizewski, Research Associate, Canadian Centre for Global Security, Ottawa, June 1993.

Dan Goodman, Oceans and Fisheries, Canadian delegate to the IWC, Government of Canada, Ottawa, April 1992 and January 1994.

Hassan Gudal, Associate Director and Secretary to AMCEN, Regional Office for Africa, UNEP, June 1993.

Jingjai Hanchanlash, Regional Director, Office for Southeast and East Asia, IDRC, September 1993.

Martin Harvey, Executive Officer, IWC, Cambridge, England, April and September 1992.

Shreesh Juyal, Associate Professor of Political Science, University of Regina, July 1993.

Hal Kane, Researcher, Worldwatch Institute, Washington, July 1993.

John Knaus, United States Commissioner to the IWC, April 1992.

Fred Knelman, Vice-President, Whistler Foundation for a Sustainable Environment, Victoria, B.C.. June 1993.

Steve Lonergan, Centre for Sustainable Regional Development, University of Victoria, June 1993.

Roger Lucy, Programme Manager for the Canadian Nuclear Safety Initiative, Department of International Trade and External Affairs, Ottawa, Canada

Robert Matthews, Chair, and Professor of Political Science, University of Toronto, August 1993.

Deidre McKay, intern, Environment and Natural Resources Division, IDRC, Ottawa, July 1993.

Craig Murphy, Professor, Department of Political Science, Wellesley College, MA, June 1993.

Donald Puchala, Director, Institute of International Studies, University of South Carolina, June 1993.

Daniel Reifsnyder, Director, Office of Global Change, Bureau of Oceans and International Environmental and Scientific Affairs, Government of the United States, Washington, July 1993.

Douglas Ross, Professor of Political Science, Simon Fraser University, June 1993.

Clyde Sanger, Director of Communications, North-South Institute, Ottawa, July 1993.

Johan Scholvinck, Chief, Population and Development Section, Population Division, Department of Economic and Social Information and Policy Analysis, United Nations, New York, June 1993.

Captain Joseph Sikes, Military Assistant, Deputy Under Secretary of Defense (Environmental Security), United States Government, Washington, July 1993.

Philippe Sands, Director, Foundation for International Environmental Law and Development, London, August, 1993.

Piyanart Supaphon, Documentalist, Mekong Documentation Centre, Mekong Secretariat, Bangkok, Thailand, August 1993.

R.B.J. Walker, Professor of Political Science, University of Victoria, June 1993.

Kazuo Watanabe, Coordinator on Environmental Affairs, PPU/DPOS, UNHCR, Geneva, June 1993.

ABOUT THE AUTHOR

Peter J. Stoett, Ph.D. (Queen's University), teaches International Relations at the University of British Columbia, Vancouver. His research has focused on global environmental problems and their relation to the international political system. He has published articles in several academic journals, including *International Journal, Coexistence, Environmental Politics, Futures,* and *Canadian Foreign Policy.* Dr. Stoett has also contributed chapters to books on development policies in southern nations, and he is currently working on a book entitled *The International Politics of Cetacean Survival,* which explores the world politics of whaling. He lives in Vancouver with his partner, Betty and their daughter, Alexandra.

SUBJECT INDEX

D

E

F

G

H

I

J

K

L

R

S

T

U

V

W

Y

Z